This
Books

OCEAN USES AND THEIR REGULATION

OCEAN USES AND THEIR REGULATION

Luc Cuyvers

Mare Nostrum Foundation

A Wiley-Interscience Publication

JOHN WILEY & SONS

New York   Chichester   Brisbane   Toronto   Singapore

Copyright © 1984 by John Wiley & Sons, Inc.

All rights reserved. Published simultaneously in Canada.

Reproduction or translation of any part of this work
beyond that permitted by Section 107 or 108 of the
1976 United States Copyright Act without the permission
of the copyright owner is unlawful. Requests for
permission or further information should be addressed to
the Permissions Department, John Wiley & Sons, Inc.

Library of Congress Cataloging in Publication Data:

Cuyvers, Luc, 1954–
   Ocean uses and their regulation.

   "A Wiley-Interscience publication."
   Bibliography: p.
   Includes index.
   1. Maritime law. 2. Marine resources conservation—
Law and legislation. I. Title.
JX4411.C89 1984   341.4'5   84-3587
ISBN 0-471-88676-9
ISBN 0-471-88675-0 (pbk)

Printed in the United States of America

10 9 8 7 6 5 4 3 2 1

Preface

The news media frequently report on a number of problems associated with our use of the sea. Some concern the depletion of valuable fishstocks or the pollution of coastal areas; others involve a vessel collision or a blowout on an offshore drilling rig. Perhaps these incidents seem unrelated to one another. After all, the poor state of some commercial fisheries in the North Sea appears to have little in common with a collision between supertankers off the African coast. However, this is not entirely the case. These incidents indicate that there may be something wrong in our approach to using the sea, and that, considering their increasing impact and frequency, the situation does not seem to be improving.

This demands our attention, for the sea will play a very important role in the future. As we need more food, minerals, and energy, we increasingly turn to the sea to obtain these resources, no longer limitless on land. We discard substantial amounts of waste in the sea, crowd it with more and larger ships, and seek to live or relax along its shores in growing numbers. This intensification of ocean use is showing side-effects: several important fishstocks have been depleted, some coastal areas are polluted, unfit for recreation or fishing. Moreover, the effects are not limited to environmental problems. There is also a chance of disagreement or even conflict, as has been demonstrated amply throughout history. These problems have not yet taken on dramatic proportions, but as pressure increases on the sea there is a possibility of inflicting irreversible damage, in which case our enjoyment of the sea and its resources will be much shorter than generally assumed.

How then to avoid this? By implementing a form of ocean management that can handle the increase in marine uses and activities. There is some form of management in effect already but, having developed piecemeal over many years in a time the oceans were still considered inexhaustible and unchangeable, it is not quite sufficient. We therefore have to come up with new rules, or at least adapt the old ones to prevailing and anticipated conditions.

Of course, all this is easier said than done. To begin, what does ocean management entail? We can define ocean management as the practice of developing and controlling marine uses and resources. This process involves a lot of data acquisition and analysis, program planning and implementation, and operational responsibilities.

Ocean management also entails a certain degree of ocean protection, not merely against oil spills and other highly visible marine pollution incidents but against overfishing, unregulated mining, too many ships, or even too many tourists. It does not imply that any of these activities should be prohibited but calls for their coordination, control, and, where necessary, regulation to prevent the misuse or waste that frequently occurs.

Some changes in our approach to using the sea are required, but before they can be implemented we need to make an inventory of our uses, new and traditional, of the sea. And this book, in the first place, intends to be such an inventory, albeit a very general one. It describes the most important marine activities and their regulation and indicates where problems will arise as they intensify. Given the immense subject matter, it unavoidably represents no more than a glance at a complex and rapidly evolving situation, but one that I hope familiarizes the reader with a number of issues that concern us all. A new approach to using the sea must command respect from all of us. One way to stimulate this respect is by providing accurate and objective information, a task I hope this study contributes to.

The book consists of eight chapters. A short introductory chapter describes some of the physical, geological, chemical, and biological aspects of the ocean, an understanding of which will be helpful throughout the text. The following four chapters review the principal uses and resources of the ocean: food, minerals, waste disposal, and navigation. Each of these is examined from a scientific as well as economic, political, and legal perspective. A sixth chapter discusses a relatively new ocean use: the development of marine energy sources. The seventh chapter focuses on the law of the sea, while the final chapter provides some conclusions and recommendations.

In examining fisheries, mineral development, waste disposal, navigation, and energy development, this study does not cover all ocean uses but some of the principal ones. Military and recreational activities, or the use of the sea for artificial islands, are among the relatively newer uses, which are expanding and adding to the intensification problem, and therefore cannot be neglected. However, because of time and space constraints, they are not included in this study.

The reader may also observe that this study, while pulling together quite a bit of information on ocean uses and their regulation, offers little in terms of concrete recommendations to improve the current situation. Since the sea is primarily international in scope, its management is a matter of international cooperation. However, we have relatively little experience or success in managing common or international resources and there are, as a consequence, few precedents on which a management regime could be based. Rather than adding to the already existing mass of lofty but impractical recommendations, I prefer to acquire more experience in areas that hinder their implementation.

Finally, the reader will undoubtedly notice that a good deal of attention is spent on the legal aspects of ocean development, the somewhat less exciting side of the coin. The body of law is of society's creation. It should reflect common sense and be capable of alteration. The sea, in contrast, is an element of nature, which may make no "sense" at all, while its complexities remain infinite. As long as law takes no or little account of the nature of natural systems, there can be no harmonious relationship between them, nor will there be effective resource management.

Many persons gave me the privilege of their advice and counsel during the research for this study. While it is impossible to acknowledge all, I would like to record my indebtedness to the following in particular: Dr. Lee G. Anderson, University of Delaware; Dr. John E. Bardach, East–West Center, Honolulu; Dr. Jean Carroz, Food and Agriculture Organization, Rome: Dr. Robert Cohen, Giannotti & Associates, Annapolis; Dr. Robert Dietz, Arizona State University; Mr. Luis Garcia, Law of the Sea Library, United Nations; Dr. P. Hovart, Fisheries Research Station, Ostend, Belgium; Dr. Clarence P. Idyll, Bethesda; Professor Gerard J. Mangone, University of Delaware; Mr. Frederick E. Naef, Lockheed Missiles & Space Company, Washington, D.C.; Dr. J. D. Nyhart, Massachusetts Institute of Technology; Dr. Arvid Pardo, University of Southern California; Dr. Guido Persoone, Ghent State University, Belgium; Mr. J. J. Saquet, Union Miniere, Brussels; Mr. Russell Stryker, Maritime Administration, Washington, D.C.; and my advisory committee at New College of the University of South Florida, where this study originated: Dr. Margaret L. Bates, Dr. Alfred Beulig, Dr. W. Tyler Estler, and Dr. Dana N. Stevens. Many thanks go to Professor Edward D. Goldberg of Scripps Institution of Oceanography, who carefully reviewed the manuscript and, where appropriate, made suggestions on how to put some rather strong terms or statements in a proper perspective. I learned quite a bit from his review.

Special recognition is due to the Stichting Leefmilieu (Environment Foundation) in Antwerp, Belgium, under whose auspices part of this study originally appeared in Dutch. Its secretary, Mr. Bavo Rombouts, was a driving force behind the study, while Dr. F. Van Zandycke, along with Lois Butler and Kathy Maas of the University of Delaware, prepared many of the figures.

Finally, very deep appreciation goes to my parents, who regularly reviewed the manuscript and offered many helpful comments as well as a good dose of encouragement. To another member of my family, this study is respectfully dedicated: my brother, Michel Cuyvers, who left us much too early and who shared a very deep love for the sea.

LUC CUYVERS

Arlington, Virginia
April 1984

Contents

OCEAN USES AND THEIR REGULATION

Chapter One

The Oceans

Most books about the sea start by pointing out that the oceans cover about ~~70 percent~~ of this planet's surface, and this study is no exception. This an ~~extremely important fact,~~ and it is indeed remarkable that greater strides have not been made to discover more about the Earth's dominant feature—but of course we were adapted to life on land, not in the sea.

Not until the last 100 years or so did we start examining the sea in earnest. Yet in this relatively short time, oceanography has made great progress, and a comprehension of some of its basic elements will be helpful in understanding what this study is about, though it mainly deals with the political and economical aspects of the oceans. Far too often misunderstandings arise among the many people involved in ocean uses, perhaps in part because the essential background needed to understand this magnificent environment was never quite acquired.

Much more is known about the oceans, of course, than can even be suggested here. Scientists have devoted lifetimes to the study of the sea. Through their efforts, the shrouds of mystery covering the sea are slowly being unraveled, exposing the amazing beauty and complexity of this planet's principal feature. But it is important to keep in mind that there are many limits on our knowledge, particularly when thinking in terms of what we can expect from the sea. Every discovery raises more questions than it answers. Our relationship with the sea contains many unknowns. It is important that these be carefully evaluated before expectations are built up, for if they are not, our enjoyment of the sea and its resources may be much shorter than often assumed.

1.1. THE STRUCTURE OF OCEAN BASINS

The world ocean has been separated conveniently into four major basins: the Atlantic, the Pacific, the Indian, and the Arctic oceans (Fig. 1.1). But this familiar pattern did not always exist. Since their initial formation, billions of years ago, the ocean basins have experienced considerable change, with the present configuration evolving only recently in geologic time.

In the early part of this century Alfred Wegener, a German meteorologist, suggested that the oceans, as we know them, were formed through the breaking up of a single supercontinent, Pangaea, at the end of the Paleozoic era, some 200 million years ago (Fig. 1.2). His theory, extrapolating boldly beyond existing data, concluded that the ocean floors were different from the continents, forming a matrix from which landmasses protruded. Following the breakup of Pangaea, he suggested, the continents drifted slowly to their present positions.

Wegener was determined to prove this theory but never was able to convince his contemporaries. In fact, it was only during the late 1950s that evidence from dated magnetic anomalies on the seafloor finally led the scientific community to endorse the concept of continental drift, or—more appropriately—seafloor spreading, completely revolutionizing contemporary geologic thought.

Figure 1.1. The major ocean basins. (From G. Dietrich, K. Kalle, W. Krauss, and G. Siedler, *General Oceanography*, 2nd ed., Wiley-Interscience, New York, 1980, Plate 1.)

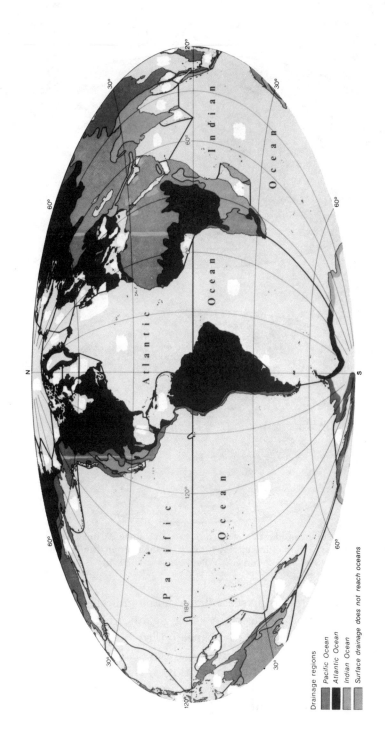

Drainage regions

Pacific Ocean
Atlantic Ocean
Indian Ocean
Surface drainage does not reach oceans

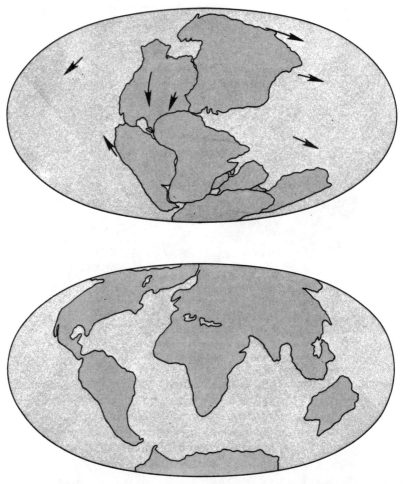

Figure 1.2.   The supercontinent Pangaea, as it appeared some 200 million years ago (top). Following its breakup, the continents slowly drifted to their current configuration (bottom). (From K. Stowe, *Ocean Science*, 2nd ed., Wiley, New York, 1983.)

The evidence supporting seafloor spreading indicates that the Earth's crust is divided into 10, or as many as 14, gigantic irregular plates, which float on the more dense material beneath (Fig. 1.3). Each plate includes oceanic as well as continental crust. New crustal material is continually formed along the axes of the oceanic ridges, and as the plates grow on either side of the ridge, they move in opposite directions, carrying te seafloor and the continents along with them. Composed of iron-rich basalt, each successive upsurge of material was magnetized in the direction of the prevailing magnetic field, which changed numerous times thoughout geologic history. It was the resultant asymmetrical polarity of upwelled crustal material on either side of the ridge that offered sufficient evidence in support of seafloor spreading.

The changes seafloor spreading has wrought on the size and shape of ocean basins are impressive. It has been determined, for instance, that the South Atlantic Ocean becomes 3 cm wider each year, while the Pacific Ocean shrinks at a somewhat faster pace. The African continent seems to be drifting northward on a collision course with Europe, inevitably closing the Mediterranean Sea. The resultant distribution of continents and oceans does not follow a regular pattern. Nearly two thirds of the land area is located in the Northern Hemisphere while the Southern Hemisphere is largely covered by water.

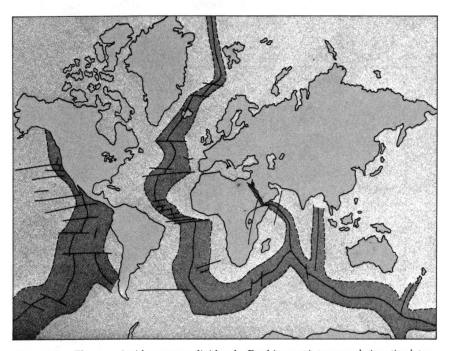

Figure 1.3.   The oceanic ridge system divides the Earth's crust into several gigantic plates. New crustal material is formed along the axes of the ridge system (solid lines); its boundaries are marked by dashed lines. (From K. Stowe, *Ocean Science*, 2nd ed., Wiley, New York, 1983.)

The average depth of the oceans is 3.7 km. This may seem substantial, but when compared to the Earth's diameter of 13,250 km, the ocean is merely a thin film of water stretched over the Earth's surface.

Some of the large-scale features of the oceans' topography are seen in Figure 1.4. The continental shelf is a relatively smooth and gently sloping extension of the continent. Much of what is now continental shelf was exposed during the last Ice Age, 10,000 years ago. Beyond the shelf, the bottom steepens to become the continental slope, rapidly dropping to depths of 3000 to 4000 m.

A large portion of the deep ocean bottom consists of flat, sediment-covered areas called abyssal plains. These are interrupted by oceanic ridges, rugged linear subsea mountain chains that encircle the globe. Isolated peaks of these mountain systems extend above the surface occasionally to form islands. Trenches, in contrast, are deep ocean depressions, generally reaching to depths beyond 6000 m. The Challenger Deep, named after the British frigate H.M.S. *Challenger*, which undertook the first full-scale investigation of the ocean more than 100 years ago, extends to a depth of 11 km in the western North Pacific—the greatest depth registered in the ocean.

1.2.   THE OCEANS IN MOTION

Ocean water is constantly in motion. Circulation processes such as tides, waves, and currents enhance mixing and minimize variations in salinity and temperature, thus providing a relatively stable medium for living organisms. In addition, circulation mechanisms serve to disperse floating organisms and their reproductive products. Pollutants and waste are

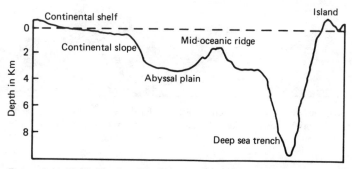

Figure 1.4.   A highly simplified cross section of the ocean bottom.

dispersed as well, however, so that pollution caused by human society has reached the most remote sections of the world ocean. An understanding of some of the large-scale oceanic transport processes is therefore important. The oceans are indeed our common heritage, rotating like soup boiling in a kettle. The spices one nation puts in will eventually be tasted by all others.

Currents

Currents have long been known and used by mariners. Some of them, such as the Gulf Stream along the western European coast and the Kuroshio off Japan, have profound effects on continental climates. Moving from equatorial to northern temperate regions, they transport large volumes of warmer waters into northern seas, resulting in milder climates for the areas that border them.

One of the major causes of this massive water transport is uneven heating. Waters at low latitudes are heated, become less dense, and spread out over the surface toward higher latitudes. As these waters drift into colder regions, they are cooled and sink. A giant convection cell is thus formed, wherein surface waters sink at the poles and flow toward the equator where they rise and spread out back toward the poles. This type of motion is called thermohaline circulation (Fig. 1.5).

Although thermohaline circulation provides for some surface water movement, it does not account for the particular pathways of the great ocean currents. Two other factors contribute to the system: the wind, itself produced by uneven heating, and the confinement of the oceans within the boundaries set by the continents.

When wind blows over water, it exerts a force on the surface in the direction of the wind. The ocean's response to this force is complicated

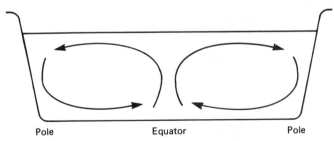

Figure 1.5.   Idealized thermohaline circulation in the ocean.

by several factors, among which the Earth's rotation is of importance. This was not generally realized until 1835 when a French mathematician, Gaspard de Coriolis, discovered a new force while studying equations of motion in a rotating frame of reference. He went on to show how the effects of the Earth's rotation could be incorporated into the Newtonian equations of motion by adding two terms: the centrifugal force of the Earth's rotation and the newly discovered Coriolis force, which makes allowance for the effects of conservation of angular momentum on a particle moving relative to the Earth's surface. The vertical component of this force is, of course, negligible in relation to the force of gravitation, but the horizontal component becomes important where forces acting along the horizontal are relatively weak, which is the case in water.

When wind blows over water, the upper layers of the surface are, as a result of the Coriolis force, transported 90 degrees to the right of the wind in the Northern Hemisphere, and 90 degrees to the left in the Southern Hemisphere, rather than being moved along the wind's direction. It is this deflection that causes ocean currents to move in a particular direction.

The wind also plays an important role in the formation of ocean currents. Prevailing atmospheric conditions can generally be accounted for by the three-celled theory. In essence, this theory proclaims that winds in the Northern Hemisphere are northeast in the latitude belt from the equator to 30 degrees, southwest in the belt from 30 degrees to 60 degrees, and again northeast from 60 degrees to the pole. A mirror image of this pattern exists in the Southern Hemisphere.

A wind blowing from the northeast causes oceanic surface layers to move northwest in the Northern Hemisphere. Similarly, the southwest winds in the area between 30 and 60 degrees cause water masses to move southeast. The combined effect of this mechanism is that water is piled up somewhere along 30 degrees north latitude, creating a high-pressure ridge in that area (Fig. 1.6a).

The horizontal pressure gradient, which is created by this pressure difference, equals the Coriolis force and produces a geostrophic flow. A parcel in the high-pressure area will tend to flow toward a region of lower pressure, but as it does so and starts to move at appreciable speeds, this motion creates a Coriolis force and the water is turned to the right. After turning 90 degrees, it cannot turn farther without flowing "uphill" and losing momentum, thus weakening the Coriolis force. If it then turns slightly to the left in response to the pressure, it regains momentum

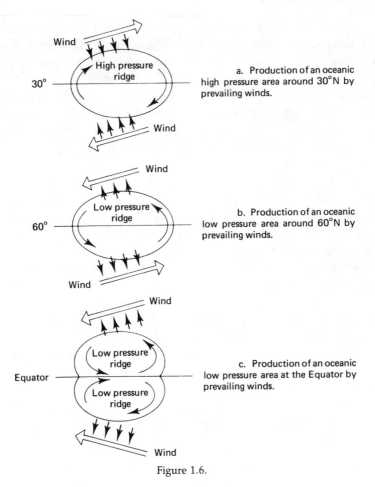

a. Production of an oceanic high pressure area around 30°N by prevailing winds.

b. Production of an oceanic low pressure area around 60°N by prevailing winds.

c. Production of an oceanic low pressure area at the Equator by prevailing winds.

Figure 1.6.

until it is forced to the right again. Eventually a balance is reached between horizontal pressure forces and the Coriolis force, with water moving endlessly around centers of high (or low) pressure. Since there are landmasses on each side of the ocean, the water tends to produce a current gyre in a clockwise direction along this high-pressure cell.

Farther north, between 60 and 90 degrees, there is a northeast wind. The same process will occur, producing a low-pressure area near 60 degrees latitude, around which waters move in a counterclockwise direction (Fig. 1.6b).

In the equatorial region, the winds produce surface layer movement away from the equator. This results in the formation of a low-pressure area with counterclockwise circulation in the Northern Hemisphere and clockwise circulation in the Southern Hemisphere. Two gyres will exist since flow direction about low-pressure areas is opposite in the two hemispheres (Fig. 1.6c).

This simple model can be compared with the surface currents existing in the oceans (Fig. 1.7). In all oceans there is a subtropical gyre, which is particularly well developed in the Northern Hemisphere, consisting of the Kuroshio system in the Pacific and the Gulf Stream system in the Atlantic. An offshoot of the Gulf Stream system, the Irminger current, combines with the East Greenland current to produce the subpolar gyre. In the northern Pacific, the subpolar gyre is not so well developed, although the Alaska current tends to develop a flow of this type. In the South Pacific, a subpolar gyre is formed by the Antarctic currents. The South Atlantic exhibits very similar properties.

In general, the model fits the real ocean quite well—much better, in fact, than would be expected from the simplicity of the initial assumptions. There are, of course, numerous discrepancies, but they can usually be explained on the basis of subsurface currents or land confinement. The proximity of the African and South American landmasses, for instance, does not permit the gyres to develop fully. As a result, there is a large transfer of water from the South Atlantic to the North Atlantic without the separation of equatorial gyres that appear in the Pacific. Finally, thermohaline effects will strengthen such currents as the Gulf Stream while weakening those opposing the drift, such as the California and Benguela currents.

Upwelling

In addition to horizontal water transport caused by currents, vertical motion is an integral part of ocean circulation. Of particular importance are the ascending motions that result in an exchange between surface and deeper waters, collectively called upwelling.

Upwelling may occur anywhere, but it is a particularly conspicuous phenomenon along the western coasts of the continents. In these areas winds run along the coastline and sweep watermasses offshore as a result of the Coriolis force. This process allows deeper, nutrient-rich waters to rise to the surface, somewhat to the discomfort of swimmers but greatly to the advantage of birds and fishermen because the water

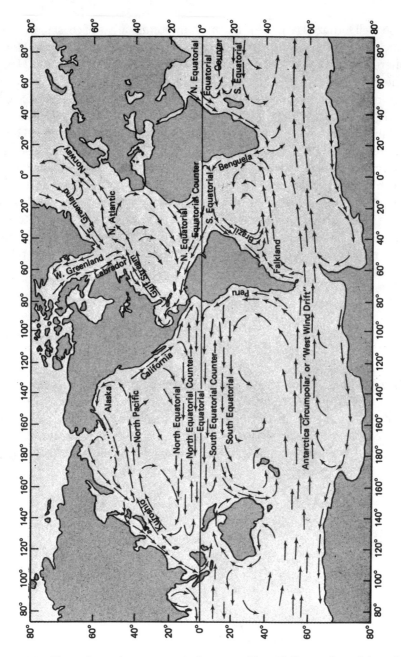

Figure 1.7. The major surface currents in the ocean. (From K. Stowe, *Ocean Science*, 2nd ed., Wiley, New York, 1983.)

is so well fertilized (Fig. 1.8). The most important areas of coastal upwelling coincide with the Canary and California currents in the Northern Hemisphere and with the Peru and Benguela currents in the Southern Hemisphere.

A different type of upwelling occurs in the Central Pacific Ocean, where the Pacific Equatorial Current flows. Here the Coriolis force causes a slight displacement of the current to the north in the Northern Hemisphere and to the south in the Southern Hemisphere. The resultant divergence of water away fromt he equator is replaced by deeper water which is rich in nutrients and sustains a relatively high biomass.

Upwelling is also observed in the Antarctic where water which sank in the Northern Hemisphere rises again, bringing with it the nutrients accumulated during its 1000-year journey. This massive upwelling permits tremendous primary productivity, which has only marginally been harvested.

1.3. THE CHEMICAL PROPERTIES OF THE OCEAN

Very few ocean processes do not involve the ocean's chemical composition. As a result, marine chemistry is very important in the study of ocean resources and their distribution. Life in the ocean, and the consequent concentration of food resources, depends to a very great extent on the distribution of chemical elements in the sea, while many marine mineral

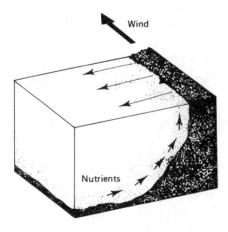

Figure 1.8. Coastal upwelling in the Southern Hemisphere.

resources are formed through complex chemical processes that can occur only in the ocean.

Seawater contains an amount of dissolved material (or salt) that is relatively large: 3.5 percent on the average. This quantity is called the salinity. Ocean circulation guarantees that this salinity is nearly constant everywhere. In addition, it has been observed that the relative proportions of the major elements remains almost constant as well.

Why the sea is salty is a question that has intrigued scientists for a very long time. One ancient theory proclaimed that heat from the sun caused the Earth to perspire, forming the oceans, which, like sweat, were naturally salty. In the many hundreds of years since this explanation was offered, chemists have derived a more plausible cause. As chemical oceanographer Wallace Broecker puts it, "The sea is a way station for the products of continental erosion. All the substances the sea receives are ultimately passed on to the underlying blanket of sediment. The great tectonic forces, that continually alter the geography of the globe, eventually bring these sediments above the sea surface and expose them to erosion. Then another trip to the sea begins." Hence the salt.

It appears that the oceans contain all the natural elements that are known. The majority of the material dissolved in sea water, however, comes from the presence of only a few elements: sodium, magnesium,

Table 1.1. **Concentration of the Major Components of Seawater**

Components	Concentration (g/kg)
Chloride	19.353
Sodium	10.760
Sulfate	2.712
Magnesium	1.294
Calcium	0.413
Potassium	0.387
Bicarbonate	0.142
Bromide	0.067
Strontium	0.008
Boron	0.004
Fluoride	0.001

Source. K. B. Turekian, *Oceans*, 2nd ed., Prentice-Hall, Englewood Cliffs, New Jersey, 1978.

calcium, potassium, chlorine, sulfur, carbon, and bromine (Table 1.1). These dissolved elements are collectively termed the major constituents, whereas all elements with smaller concentrations are classified as minor constituents.

The most dramatic variations in concentration are shown not by the major components but by the so-called nutrient elements: phosphorus, silicon, and nitrogen. Along with oxygen, hydrogen, and carbon, they constitute the building blocks of life, being converted into organic compounds by marine plants and reorganized in the bodies of marine animals.

Nitrogen is perhaps the most important nutrient element, since it is a component of the amino acids, which are necessary for protein synthesis. The main source of combined nitrogen in the sea is nitrate (NO_3^-), which may become depleted in the upper layers of the sea by the photosynthetic action of plants. Phosphorus, found in even smaller concentrations, is also extracted by phytoplankton. Both elements can be limiting factors to the production of marine life since primary production comes to a halt as they become depleted in the surface waters.

Nutrient elements are released to deeper water as a result of progressive metabolism. In areas where upwelling occurs, the nutrients are recycled to the surface almost immediately. Because of this virtually unlimited nutrient supply, these regions sustain a very high primary productivity, which attracts populations of higher life forms such as valuable fish stocks. In most parts of the ocean, this nutrient exchange does not occur quite as rapidly and the nutrients are depleted, ofttimes limiting primary production and the associated abundance of fish stocks.

The concentrations of many other elements vary as well. Trace elements, shell-forming components, and, of course, oxygen are needed for the production of life, and their concentrations are biologically controlled.

Marine chemistry is of vital importance with respect to more than just the distribution of living marine resources. Profitable recovery of a number of elements contained in seawater is, or may become, feasible and the recovery of minerals from the ocean bottom is considered to have great potential. In both instances, the development of resource exploitation techniques will involve considerable chemical technology.

In addition, marine chemical research may offer some clues to the ability of the ocean to accept society's waste. The identification of processes and substances that deleteriously affect the marine environment should be of very high priority, though in many instances this type of research is conducted only after the harm has been inflicted.

1.4.   LIFE IN THE OCEAN

It is generally agreed that life originated in the ocean. The sea offered advantages for the production of life over other zones, including protection from excessive amounts of ultraviolet radiation and a relatively stable environment in terms of temperature and the concentration of major chemical elements.

Like life on land, plants and animals in the sea exhibit an enormous diversity in size, form, and function. The range in size taken by itself is truly impressive: from microscopic one-celled plants and bacteria to the great blue whale, apparently the largest animal ever to live on earth. Yet even more impressive is the range in form and activity. Several systems of organizing animals and plants in groups have been developed to classify this enormous diversity.

Through the evolutionary processes that have operated for the past billion years, each of the many species of living organisms exhibits some genetic relationship to all others. Taxonomic classification takes account of these evolutionary relationships. Its purpose is to categorize plants and animals into natural units by tracing the lines of their evolution and by identifying similarities among them.

The fundamental and smallest unit of taxonomic classification is the species. Individual organisms classified as a species are judged to play a discrete or, at least, definable role in the living world. Members of the same species form a homogeneous genetic unit: they represent a group of closely related organisms that can and normally do interbreed, forming a fertile offspring. This exchange of genetic information steers individuals of such groups along a common evolutionary path.

The next step in the taxonomic hierarchy is conceived by grouping as a genus those species which are judged to have a common ancestor. Genera are grouped into families, which, in turn, are organized into classes and orders until finally all living things are classified in some 30 plant and animal phyla. Table 1.2 presents a taxonomic breakdown of some well-known marine organisms.

Another way to classify marine organisms is according to where they live (Fig. 1.9). The benthos includes animals living on the bottom (epifauna) or in the sediment (infauna). Benthic organisms exhibit a great variety of habits. Some animals such as oysters and barnacles attach themselves to a hard substrate. Others secure themselves to the substrate by means of a rootlike structure. Several familiar invertebrates simply lie unattached

Table 1.2. A Taxonomic Breakdown of Some Well-Known Marine Organisms

Species	Kingdom	Phylum	Class	Order	Family	Family	Genus
Eastern oyster	Animalia	Mollusca	Pelecypoda	Ostreoida	Ostreidae	*Crassostrea*	*virginica*
Octopus	Animalia	Mollusca	Cephalopoda	Dibranchia	Octopodae	*Octopus*	*vulgaris*
Grey shrimp	Animalia	Arthropoda	Crustacea	Caridea	Crangonoidae	*Crangon*	*crangon*
Purple sea urchin	Animalia	Echinodermata	Echinoidea	Echinoida	Strongylocentrotidae	*Strongylocentrotus*	*purpuratus*
Bull shark	Animalia	Chordata	Chondrichthyes	Squaliformes	Carcharhinidae	*Carcharhinus*	*leucas*
Herring	Animalia	Chordata	Osteichthyes	Clupeiformes	Clupidae	*Clupea*	*harengus*
Atlantic cod	Animalia	Chordata	Osteichthyes	Gadiformes	Gadidae	*Gadus*	*morhua*
Blue whale	Animalia	Chordata	Mammalia	Mysticeti	Balaenopteridae	*Balaenoptera*	*musculus*
Dolphin	Animalia	Chordata	Mammalia	Odontoceti	Delphinidae	*Delphinus*	*delphis*
Kelp	Plantae	Phaeophyta	Phaeophycae	Laminariales	Lessoniaceae	*Macrocystis*	*pyrifera*

on the bottom or crawl and propel themselves along the surface of the seabed.

Pelagic organisms, in contrast, inhabit the water column. They are generally classified into two groups: plankton and nekton. Planktonic organisms, their name derived from the Greek work for wanderer, maintain a specific gravity very close to that of seawater. Since they have little or no capability of horizontal motion, they are carried along with the current. Plant members of the plankton are called phytoplankton. They are mostly microscopic, either single-celled or loose aggregates of a few cells. The animal plankton is referred to as zooplankton. Its members range in size and complexity from single-celled organisms to multicellular animals. Some of the larger planktonic animals have remarkable vertical swimming abilities, but most display the passive, floating habit characteristic of planktonic life forms. The large, actively swimming marine animals belong to the nekton. This group includes all marine mammals, many fish, and a few invertebrates such as squid and some shrimp.

In many instances, however, these clear-cut distinctions between groups of organisms living under similar environmental conditions breaks down. Some species occupy only one habitat over their entire life span, but many fish, for instance, change their mode of life systematically during their life cycle, developing from a temporary planktonic larval stage to nektonic animals as their size and swimming abilities increase.

Finally, plants and animals can also be classified by their trophic associations. Basically this approach involves an analysis of different paths of energy capture or, in simpler terms, what an organism eats. All living organisms require matter and energy from their nourishment.

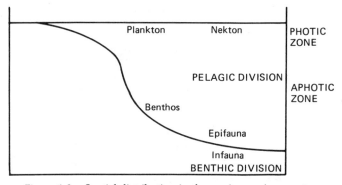

Figure 1.9.   Spatial distribution in the marine environment.

Matter is needed for individual growth and reproduction, while energy is necessary to perform work and to maintain chemical order.

Two major types of energy capture exist: autotrophy and heterotrophy. All plants are autotrophic, or self-nourishing. They do not depend on other organisms for food; instead they take inorganic compounds such as water, carbon dioxide, and nutrients to produce more complex organic compounds. A variety of energy sources are used to carry out this synthesis. The vast majority of autotrophs use readily available sunlight (photosynthesis), but other systems exist. Chemotrophic organisms, for instance, derive energy from the reduction of nitrates or sulfates (Table 1.3). Heterotrophs, in contrast, are unable to produce their own food from inorganic substances and depend on plants for nourishment. The organic matter in their food provides the chemical energy to carry out metabolic activities.

All life on Earth, including life in the sea, depends on this flow of energy which originates in the sun, enters the biosphere through photosynthetic activities of green plants, and is transferred from one organism to another in chemical form as food. In this system, plants are referred to as primary producers, which places them in the first trophic level. All higher trophic levels are occupied by heterotrophs: animals adapted to feed on plants are herbivores, forming the second trophic level, while carnivores occupy the third and higher levels. The decomposers, primarily bacteria and fungi, exist on particles of organic matter and on the remains of other organisms (Fig. 1.10).

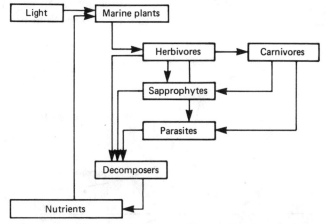

Figure 1.10.   A simplified model of energy flows in the marine environment.

Table 1.3.   **Metabolic Reactions Occurring in the Oceans**

Oxidation by oxygen:
$$(CH_2O)_{106}(NH_3)_{16}H_3PO_4 + 138\ O_2 \rightarrow 106\ CO_2 + 122\ H_2O +$$
$$16\ HNO_3 + H_3PO_4$$
Denitrification:
$$(CH_2O)_{106}(NH_3)_{16}H_3PO_4 + 94.4\ HNO_3 \rightarrow 106\ CO_2 + 55.2\ N_2 +$$
$$177.2\ H_2O + H_3PO_4$$
Sulfate reduction:
$$(CH_2O)_{106}(NH_3)_{16}H_3PO_4 + 53\ H_2SO_4 \rightarrow 106\ CO_2 + 53\ H_2S +$$
$$H_2O + 16\ NH_3$$

Living organisms, like most energy-consuming systems, are not very efficient in their use of energy. Since each organism uses part of the available energy for maintenance, growth, and reproduction, a loss of energy from one trophic level to the next is inevitable. A widely quoted value for the efficiency of energy transfer is 10 percent, though many instances exist where they are significantly higher, or even lower. All this is very important, as it indicates that on the average only one tenth of the energy represented by one level is assimilated by or available in the next.

The paths of nutrient and energy flow through the living portion of the marine ecosystem are called food chains or food webs. With few exceptions, the first level of the marine food chain is occupied by widely dispersed microscopic plants. Their microscopic character imposes a size restriction on the occupants of higher trophic levels, since most animals are not adapted to feed on organisms many orders of magnitude smaller than themselves. As a result, herbivores in the sea, in contrast to on land, tend to be small, and large marine animals are usually carnivores, occupying levels far up the food chain.

This chapter could continue, reflecting its principal message that life in the sea is complex, forming part of an intricate system about which we know relatively little. Perhaps it is safe to state that we have achieved a certain level of penetration, of taking things apart, in our knowledge of the sea, but before us lies the task of comprehension, of putting the pieces together. It is exactly this type of understanding that is needed as a basis for sound management.

Chapter Two

Food from the Sea

Archeological evidence in the form of shellmounds, fishhooks, and fishbones indicates that the sea has provided food for as long as people have lived along its shores. In early classical times, the Egyptians, Phoenicians, and Carthaginians relied on the fisheries of the Mediterranean, as many ancient temple drawings show, and at the height of the Roman Empire, Rome's colonies supplied the Imperial City with preserved seafood of many varieties, including tuna, swordfish, sturgeon, eels, oysters, and sea urchins. During the Middle Ages, this reliance on the sea's food resources increased, particularly in coastal areas but also far inland as a result of the development of better preservation techniques.

Thousands of years after people started to fish, they are still doing so as hunters. Techniques and efficiency have been drastically improved in the course of time, however, and what once seemed like an inexhaustible source of food has been depleted in some instances to very low levels. We expect the sea to play a major role in providing food to an ever-increasing world population, but if this is to be the case, some important changes need to be made in the management of the sea's living resources.

2.1. FISH AS A FOOD SOURCE

At the end of the eighteenth century, Thomas Robert Malthus, a British economist, asserted that man would not be able to produce food fast enough to keep up with population growth. Population, he wrote, increases geometrically, food production merely arithmetically and, as a result, mankind is doomed to famine and poverty. Malthus was widely criticized when he published these rather disturbing theories, yet he was one of the first scientists to perceive the problems accompanying the tragic increases in the world's population. It has in the meantime become clear that population growth is indeed geometrical: in 1650 there were 500 million people in the world, in 1830 one billion, in 1930 two billion, in 1960 three billion and at present well over four billion—in fact, more than 4.6 billion as of late 1982. The world population currently increases by more than 2 percent each year, which means that every year there are 80 million additional mouths to be fed.

Certainly, food production has not been able to keep up with this growth, increasing only about 0.5 percent each year. Food shortages have been the result. Millions of people die each year of starvation or diseases exacerbated by starvation. There is little that can be done about this, unless food supplies keep up with population growth. Famine is perhaps society's most alarming problem, but it is based on one of nature's essential laws: the carrying capacity of a certain area is limited to its food-producing capacity. Efforts to feed the famished may not solve the problem but can defer the starvation of far greater numbers.

The sea has always been regarded as offering great potential in the quest for food. This belief found firm support during the 1950s and 1960s when fish catches rose spectacularly, sometimes as much as 6 percent annually, which was much more than any other food source. This increased flow of food was, of course, perceived with much optimism. In addition, little was known at that time about food production in the sea, and its sheer size, coupled with this lack of information, led to the belief that the sea's food resources were inexhaustible. Noted historian Arnold

Toynbee summed widely held feelings when he stated that the sea was "a vast accessible field for mankind's enterprise, and also a sure guarantee for our race's survival even if our descendants are going to be ten times as numerous as we are today." Popular accounts abounded on the potential of farming the sea by cultivating edible seaweeds and by breeding or herding fish, or the possibility of plankton harvests as an economic and acceptable source of protein, contributing to some misconceptions that still prevail today.

Table 2.1. **Marine Fisheries Catch, 1970–1980 (metric tons per country)**

Country	1970	1975	1980
Japan	9,182,100	9,696,449	10,189,656
USSR	6,390,600	9,026,010	8,664,775
USA	2,810,800	2,764,803	3,564,954
People's Republic of China	3,000,000	3,182,309	3,000,000
Chili	1,228,300	899,458	2,816,614
Peru	12,532,900	3,440,861	2,719,695
Norway	2,985,700	2,481,473	2,398,171
South Korea	842,100	1,878,355	2,052,023
Denmark	1,217,100	1,750,633	2,010,481
India	1,085,600	1,482,105	1,548,155
Iceland	733,300	994,275	1,514,376
Thailand	1,343,400	1,392,144	1,500,000
Indonesia	807,200	988,430	1,414,097
North Korea	1,00,000	1,000,000	1,330,000
Canada	1,344,700	950,677	1,252,319
Spain	1,527,000	1,497,358	1,207,100
Philippines	844,200	1,228,806	1,135,231
Vietnam	668,300	837,200	837,200
France	782,500	784,495	765,393
Brazil	432,800	579,536	680,000
All others	12,452,100	12,345,823	13,975,960
World total	61,982,400	59,171,200	64,576,200

Source. FAO, *Yearbook of Fishery Statistics*, Volume 50, Rome, 1981.
Note. Landings did not increase much during the 1970s—the result of overfishing and depletion.

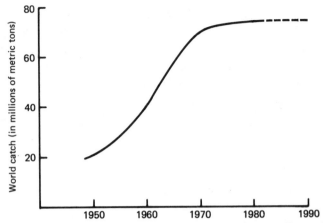

Figure 2.1.   The world fish catch (including freshwater species) in millions of metric tons during the period 1950–1980. During the early 1970s the harvest started leveling off at around 70–75 million tons and has remained there ever since.

At present, the sea annually provides some 65 million metric tons of food, almost half of which is taken by a few nations (Table 2.1). The sea's food harvest has remained almost stable since the early 1970s, definitively proving the spectacular increases in the previous years a thing of the past (Fig. 2.1). The contribution of different species to total world fish landings is shown in Table 2.2, while Table 2.3 indicates the share of each of the major ocean areas.

Of this total of 65 million tons, about one third is used for fishmeal production and does not reach people directly. Despite this roundabout route in the foodchain, conservative estimates indicate that 4 to 5 percent of our total protein consumption is derived from the sea. In addition, fish are very nutritious. Their protein indeed contains an excellent combination of the essential amino acids, which cannot be synthesized by most animals and must be supplied already manufactured. The high quality of the sea's protein yield makes it even more important. Whether this supply can be tapped further is therefore a crucial question in the search for additional food sources.

2.2.   THE FOOD POTENTIAL OF THE OCEAN

Many of the early estimates of the ocean's food potential were based on intuition and hope rather than sound scientific data. Rather optimistically,

Table 2.2.   **World Fish Landings (1980) by Groups of Species**

Species Group	Landings
Herrings, sardines, anchovies, etc.	16,225,200
Cods, hakes, haddocks, etc.	10,719,675
Jacks, mullets, sauries, etc.	7,338,318
Redfishes, basses, congers, etc.	5,247,227
Mackerels, snoeks, cutlassfishes, etc.	4,226,312
Tunas, bonitos, billfishes, etc.	2,489,795
Flounders, halibuts, soles, etc.	1,084,367
Miscellaneous marine fishes	7,581,510
Shads, milkfishes, etc.	817,990
Salmons, trouts, smelts, etc.	770,276
Eels	91,636
Miscellaneous diadromous fishes	125,295
Sea spiders, crabs, etc.	842,256
Lobsters	108,134
Squat lobsters, nephrops	56,227
Shrimps, prawns, etc.	1,680,954
Krill	424,821
Miscellaneous marine crustaceans	66,115
Abalones, winkles, conchs, etc.	86,595
Oysters	972,885
Mussels	613,965
Scallops, pecten, etc.	364,173
Clams, cockles, arkshells, etc.	1,176,171
Squids, cuttlefishes, octopuses, etc.	1,572,098
Miscellaneous marine molluscs	165,231

Source.   Adapted from FAO, 1981.

it was observed that the ocean covers an area more than twice that of land, and that food production in it is three-dimensional. But no mention was made of the fact that the ocean environment is very different for the land, with smallness, mobility, and long food chains prevailing in the ocean, in contrast with the production of life on land.

Estimates of the ocean's food potential require scientific data, and there are two methods for obtaining these. The first involves a global estimate of the production of living matter and a calculation of the amount of fish that would normally be available to people by considering the number of trophic levels and their efficiencies. The second method requires individual estimates of the sizes of various commercial fish stocks, which can then be totaled.

Table 2.3.   **1980 Fishery Landings of the Major Ocean Areas**

Area	Catch (metric tons)
Northern Atlantic	14,608,502
Central Atlantic	6,901,691
Southern Atlantic	3,864,881
Indian Ocean	3,750,887
Northern Pacific	20,730,370
Central Pacific	8,125,787
Southern Pacific	6,594,109
Total marine areas	64,576,200

Source. Adapted from FAO, Rome, 1981.

In the first approach, information is needed on global primary productivity, that is, the rates of production of phytoplankton (or primary producers) on which fish and all other marine animals ultimately depend. Several methods exist to accomplish this; the simplest way is to let the algae settle from a water sample onto the base of a special capsule, and to count them from underneath with an inverted microscope. This is a cumbersome task, however, and all the algae may not be counted because some of the smaller flagellates do not sink very rapidly and may disintegrate upon preservation.

Another way of estimating productivity is by determining the quantity of chlorophyll in a water sample. The water is filtered and the residue is dissolved in a 90 percent acetone solution. The amount of chlorophyll can then be estimated by spectrophotometry. This technique usually provides an overestimate, however, since in addition to chlorophyll it also detects its breakdown products, which are free in the water as phaeophytins or chlorophyllides.

A more accurate and direct method is the radiocarbon technique, whereby an increment of algal carbon is measured during a period of several hours. Of all methods estimating primary production, the radiocarbon technique is the most reliable because it includes those flagellates inaccessible to counting and is not vulnerable to biases as a result of degradation products.

These techniques have been used in many studies undertaken to determine the amounts of organic material produced in the sea. As can be expected, primary production is not uniform throughout the ocean since it depends on the amount of sunlight, the concentration of nutrients, the temperature, atmospheric conditions, and a variety of other factors that tend to differ from one area to another. As a result, productivity varies markedly at different latitudes as well as throughout the year. This variation is graphically represented in production cycle diagrams, which are derived by making a number of production estimates in a particular region at different times of the year.

Usually three types of production cycles are distinguished, corresponding to polar, temperate, and tropical regions (Fig. 2.2). The polar cycle is characterized by a high amplitude of short duration. Temperate cycles consist of an early spring production peak, followed by a smaller autumn peak. Tropical cycles, on the contrary, show relatively continuous production at low amplitudes. These diagrams clearly indicate that production is by no means uniform throughout the ocean. Instead, most of the ocean is a virtual desert, represented by the low-amplitude production cycle of the tropics. It may seem somewhat contradictory that these areas, with plenty of sunlight and clear waters, are the least productive, but, as noted earlier, nutrient availability may be the limiting factor in primary production, and in the tropics nitrates and phosphates are in continuously low supply. This is not the case in polar regions where divergencies, particularly in the Antarctic, supply massive amounts of nutrients to the surface, permitting a very high production during the summer months when sunlight is present. Nutrient availability usually is not continuous in temperate regions, which explains the occurrence of two peaks. It is assumed that the autumn peak is initiated through the recycling of nutrients that became depleted as the spring peak progressed.

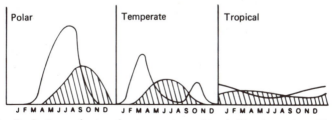

Figure 2.2. Production cycles in polar, temperate, and tropical marine regions. The vertical axes indicate production rates, usually measured in grams of carbon per unit surface area.

Table 2.4. **Primary Production in the Sea**

Region	Percentage of Ocean	Area (km²)	Mean Productivity (g carbon per year)	Total Production (billions of tons of carbon per year)
Oceanic	90	326×10^6	50	16.3
Coastal	9.9	36×10^6	100	3.6
Upwelling	0.1	3.6×10^6	300	0.1
Total				20.0

Source. Adapted from Ryther, 1969.

Depending on phytoplankton for food are the herbivores, the second trophic level. They reach their highest amplitude shortly after the primary production peak (Fig. 2.2). The time lapse between the onset of effective grazing and the production peak is called the delay period. High latitudes characteristically have long delay periods, since it takes a new generation of herbivores to start effective grazing. In polar waters, and to a large extent in temperate regions, herbivores have to live part of the year with little food, usually in deeper waters. As a result, the animals tend to be bigger, carry larger fat stores, and have longer life cycles than their counterparts in the tropics. Because of this, food chains in polar areas are shorter than those in tropical areas, making temperate and polar waters more productive in terms of harvestable biomass.

Finally, although the metabolic loss between successive trophic levels is classically assumed to be 90 percent, ecological efficiencies—that is,

Table 2.5. **Fish Production in the Ocean**

Area	Primary Production (tons carbon per year)	Number of Trophic Levels	Efficiency	Fish Production (tons)
Oceanic	16.3×10^9	5	10	16×10^5
Coastal	3.6×10^9	3	15	12×10^7
Upwelling	0.1×10^9	1.5	20	12×10^7
Total				24×10^7

Source. Adapted from Ryther, 1969.

Table 2.6. **Estimates of the Food Potential of the Ocean**

Author	Potential (millions of tons)	Year
Chapman	1000	1966
Pike and Spilhaus	200	1965
Ryther	100–120	1969
Ricker	150	1968
Moiseev	80–100	1964
Cushing	100	1966
Bogorov	100	1965
Schaefer	200	1965
FAO	120	1969

Source. Adapted from Moiseev, 1969.

the efficiency of food transfer—can be quite different. In tropical areas food is small and dispersed and, as a result, transferred at higher efficiencies than is the case in northern regions, where it is abundant and passed through the system at considerably lower efficiencies. But food conversion is also affected by its availability: the more available the food, the less work the animal must perform to obtain it. Tropical herbivores, although "internally" more efficient, have higher "external" metabolic costs than polar herbivores. The two metabolic costs work in opposite ways, more or less toward a constant total effect.

To assess the sea's food potential, estimates are needed for all these values and parameters. First, global primary productivity needs to be derived. This can be done by dividing the oceans into a number of regions and assigning values to the total area and the average productivity in each region (Table 2.4). The next step involves some more arbitrary estimates, because in order to arrive at a total biomass, values need to be assigned to the number of trophic levels and the efficiency of food transfer, parameters which cannot really be generalized (Table 2.5). Using this method, John Ryther, a biologist at the Woods Hole Oceanographic Institution, obtained a total of 240 million metric tons of fish in the sea, which would allow for a sustainable yield of about 120 million metric tons. There is no doubt that this represents an educated guess at best, thought it does not differ considerably from other estimates (Table 2.6).

The ocean's food potential can also be estimated on the basis of individual stock size assessments. To make a relatively accurate estimate of the size of a fish stock, numerous data are needed. Fortunately, a lot of this information is readily available since up to half of well-exploited fish stocks lands in fish markets, where fishery biologists can determine the age and the length of a fraction of the catch and obtain the fishermen's findings on the abundance and location of the stock. This information, in turn, is plotted in models that represent the events that occur in fish stocks, yielding estimates on their size and potential catch.

Any population of animals is maintained by the balance between its birth and death rates. The birth rate is determined mainly by natural selection, while the death rate reflects the capacity of the environment to carry the population. Common to all (unexploited) animal populations is a decline in mortality during juvenile and immature life, with the smallest death rate occurring in early and middle mature age groups and increasing in old age groups. As fish are very fecund (the number of eggs per female can reach the millions), early mortalities are extremely high: from 5 to 10 percent per day, a figure which falls to no more than 20 percent per year in early and middle maturity. Fish grow rapidly during juvenile life, more slowly with maturation, and very slowly in old age. They grow continuously, however, and even during the period of mature life they may increase several times in size.

A fish stock is, generally speaking, at its most abundant when it is not exploited. In this stage, it includes a relatively high proportion of large and older individuals. Each year a number of young fish enter the stock—these fish are called recruits—and all the fish put on weight. This increase in growth and population, in turn, is balanced by natural mortality.

When the stock becomes exploited, the population structure changes considerably. Fishing represents a new mortality factor, which will reduce the level of the stock progressively. At reduced population levels, the losses accountable to natural death will be less than the gains resulting from recruitment and individual growth. If the catch is less than the difference between natural losses and gains, the stock will increase again. If the catch is more, however, it will decrease. When the stock is not decreasing or increasing, the catch is equal to surplus growth and a sustainable yield is obtained. The sustainable yield is small when the stock is large and when it is small. It reaches a maximum when the stock is at an intermediate level—somewhere between one third and two thirds of the unexploited population (Fig. 2.3). In this intermediate range

Yield

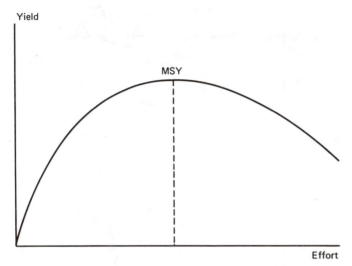

MSY

Effort

Figure 2.3.   The sustainable yield curve.

the fish are younger and smaller than in the unexploited condition, but individual growth is higher.

While this is true for all fish populations, a detailed analysis of each fishery is needed to determine how much and what size fish should be caught each year to obtain a sustainable yield. This is where the models are used. Fishery models can be grouped conveniently in two groups: those that treat the fish population as a whole, considering the changes in biomass without referring to its structure, and those that look at the population as the sum of its individuals and are concerned with the growth and mortality rates of individual fish.

The first type of model is called a descriptive model. It is based, in essence, on the hypothesis that the relative rate of growth of an unexploited population is a monotonically decreasing function of its own biomass. The simple rationale behind this hypothesis implies that the environment in which the population is found has a limited capacity for supporting it. For small populations, that capacity is not fully utilized, but as the population increases, more and more of it is used for maintenance, while competition among individuals becomes keener. In due course, the population reaches a level where all of the food supply is needed for maintenance, and no surplus growth takes place. This defines the natural equilibrium level of the stock (Fig. 2.4).

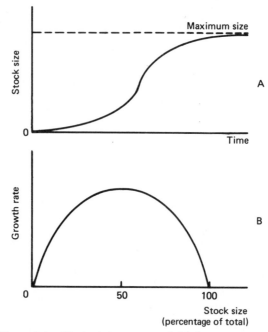

Figure 2.4.   The logistics curve and the descriptive model.

The relation between the growth rate and the population size is described by the logistics curve, which states that the net rate of increase, from generation to generation, is balanced by a constant, assumed to be the carrying capacity of the environment. The rate of increase is determined by the current size of the population. At zero and equilibrium populations, the stock will not change. For any population between, the rate of increase is positive. The increase will rise to a maximum at some internal value and then decrease, whereafter the stock asymptotically approaches the maximum population size (Fig. 2.4).

The curve in Figure 2.4, even though an extremely simplified representation of the changes in a fish stock under exploitation, illustrates some important biological features of a fish stock. First, it is impossible to exploit a fish stock without effecting some change. This may indeed seem obvious, but with the present-day desire to minimize ecological impacts, there may be a feeling that a well-managed fishery is one that causes no change, and this is simply impossible. The second point was mentioned earlier: provided catches are not too great, the decline in population is not continuous. After some time the population will reach a new equilibrium, from which the same amount can be removed year after year without affecting the stock.

To determine these yields, and in particular the maximum sustainable yield, fishery scientists need data on the size of the stock, the catch, and the effort applied to obtain it. In practical fishery situations, the biomass of a stock can rarely be measured directly, so more readily available information, such as catch and effort, is relied upon. Provided that there is enough information, scientists can relate catch to stock and effort and thus determine sustainable yields. Observations are required annually, however, and it may take 10 years or so to pinpoint the position of the maximum.

Moreover, although the descriptive model requires only the simplest types of data, there is no proof that fish stocks follow the model's basic assumptions. As a result, this method can be considered sufficiently reliable only for stocks that cannot easily be aged. Perhaps its most valuable application was to establish a maximum sustainable yield for the Antarctic blue whale: the simple presentation of the facts eventually convinced the International Whaling Commission that conservation was needed, and needed quickly.

A more accurate and rapid method to assess potential stock yields is the analytic model, first formulated as follows:

$$P_2 = P_1 + (A + G) - (C + M)$$

in which P_1 is the size of the stock at the beginning and P_2 at the end of the year, A is recruitment, G is growth, C is catch, and M represents natural mortality. If the stock is in equilibrium ($P_1 = P_2$), then

$$A + G = C + M \quad \text{or} \quad C = A + G - M$$

This simply states the factors that govern catch and stock size. The population is considered as the result of changes in these parameters in the analytic model, rather than as a series of increments in weight or number, as in the descriptive model.

This approach was used by Beverton and Holt, two British population biologists, who considered the stock as the integral of numbers and of weight of individual fish from the age of recruitment to the age of extinction in a year-class. In the simpler models, the rate of growth, natural mortality, and recruitment are assumed to be constant and independent of the abundance of the stock or the amount of fishing. Normally, it is also assumed that individual fish of the same age do not differ with respect to their rate of growth or their susceptibility to capture. These assumptions can be adjusted in more realistic models, as the situation demands and as more data become available.

The Beverton and Holt analysis begins by considering the history of a group of recruits throughout their lives after they enter the stock at a catchable age. During any period of time, a fish may be caught, die of natural causes, or survive to the beginning of the next period. The survivors can be calculated as the difference between the number alive at the beginning of the period and the number of deaths during the period. The number of deaths is a function of fishing and natural mortality. The catch in weight is equal to the product of the numbers caught and their average weight, which can be obtained easily using growth curves or tables of weight at different times in the fish's life cycle. The total catch, in turn, is obtained by addition of catches in the individual time intervals.

Several parameters need to be determined in the analytic model. Natural mortality is difficult to estimate. Usually only the total mortality is easily obtained and a variety of techniques can be used to separate fishing from natural mortality. Parameters relating to growth are easily obtained if the age of the fish can be determined without too much trouble, which is usually the case. The age of recruitment can be estimated by using patterns of growth as well as some knowledge of the general biology and behavior of the fish. The most difficult parameter to estimate is recruitment, because in many fisheries annual recruitment fluctuates widely and apparently at random. Consequently, yield is not expressed as catch but as yield per recruit. With constant recruitment, the maximum yield per recruit will maximize the catch on a sustainable basis (Fig. 2.5). Even though predicting recruitment remains the main difficulty in the analytic model, the scientific advance made by Beverton and Holt is considerable in that they explored the age structure of the stock as shown in the market sampling system. They devised models based on population parameters which provide the information needed for management purposes rapidly, rather than after an extended period of observation.

A step beyond the descriptive and analytic models, both called single-species models, is the multispecies model. Though it is not described here, it deserves mention because the exploitation of one stock nearly always affects one or more others. Multispecies models take account of this interdependency. As should be the case, fishery biologists use these models increasingly to predict the effects of fishing on a stock and to

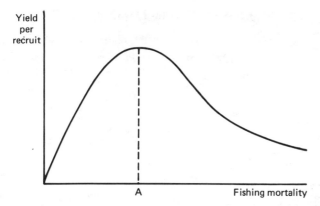

Figure 2.5.   The analytical model: yield per recruit and instantaneous fishing mortality (with constant recruitment). Fishing pressure in excess of A leads to overfishing.

give advice on the amounts that can be removed without harming it.

Fishery models permit scientists to make estimates of the potential yield of individual fish stocks. The sum of the yields of all stocks in the ocean also gives an indication of the ocean's food potential (Table 2.7). Most such estimates show yields in the 100 million to 120 million metric ton range, less than double our current harvest. It must be noted, however, that such yields cannot be attained unless presently overexploited stocks are given a chance to recuperate.

In addition to conventional supplies of marine protein, there also exists a possibility of increasing the ocean's food yield by exploiting unconventional resources. Among these, greatest progress has been made with krill, a shrimplike organism extremely abundant in the Antarctic Ocean (Fig. 2.6). Currently, between 700,000 and 1 million metric tons of krill are caught each year, but estimates on its potential catch range anywhere from 10 to 150 million metric tons, more than double the present total world catch. Technically efficient methods for its harvest have been developed, but the scale of an operation would have to be very large to provide a product competitive with other forms of animal protein, which could entail producing krill in quantities greater than most markets could absorb. Krill also decomposes very rapidly and must be processed immediately after capture. The Japanese and the Russians have developed a technique whereby krill is processed into a paste, similar in flavor to shrimp. Krill have also been presented on the Japanese market in dried form. Though the cost of processing and deep freezing

Table 2.7. **The Food Potential of the Ocean through Individual Stock Size Assessments**

Region	1975 Catch (millions of tons)	Percentage of Potential	Total Potential (millions of tons)
Atlantic Ocean			
Northwest	3.8	55.9	6.8
Northeast	12.1	86.4	14.0
Western central	1.6	27.6	5.8
Eastern central	3.5	97.2	3.6
Southwest	0.9	11.7	7.7
Southeast	2.6	56.5	4.6
Mediterranean and Black seas	1.3	100.0	1.3
Total	25.8	58.9	43.8
Indian Ocean			
West	2.0	21.5	9.3
East	1.1	19.6	5.6
Total	3.1	20.8	14.9
Pacific Ocean			
Northwest	17.0 ⎫	89.2	27.1
Northeast	2.2 ⎭		
Western central	5.0		
Eastern central	1.3	20.3	6.4
Southwest	0.3	50.0	0.6
Southeast	4.6	35.1	13.1
Total	30.4	64.4	47.2
Total	59.3	56.0	105.9

Source. FAO, *Review of the State of Exploitation of World Fish Resources*, Committee on Fisheries, Rome, 1977. (Prepared by J. A. Gulland.)

the product for direct human consumption appears to be decreasing, a significant demand for it does not as yet exist.

Other Southeast Asian nations and some European countries, notably Poland and Germany, are involved in krill exploitation. At present, these activities are not regulated since the amounts taken are insignificant, but with the possibility of mass exploitation ahead, a management regime will become a necessity. A treaty to permit the catch and management of this resource was opened for signature in May 1980 but may not be enforced for several years. The treaty takes account of krill's dominant position in the Antarctic ecosystem but its efficacy remains, of course, to be seen.

Somewhat farther away is the utilization of lanternfish, small, bony mesopelagic fish that are widely distributed throughout the oceans. The resource potential of this species has been estimated at 100 million metric tons, again an immense amount. There are quite a few technical problems with respect to the catching of such a widely dispersed fish, however, and it appears that even greater difficulties will be encountered in the development and marketing of a suitable low-cost product. If exploited, lanternfish may therefore be processed into fishmeal. In the long term it is possible that the stocks could directly contribute to human consumption, particularly since they are located in tropical regions, near areas with nutritional problems.

Of equal potential, and similarly underused, are large stocks of cephalopods, particularly squid. As in the case of lanternfish, extremely large catches of squid are possible, yet mass production remains a problem. In both instances, techniques are needed to harvest the resources within the boundaries of economic reality, and additional markets need to be created. It may seem cynical that a world suffering from a lack of protein would have difficulties marketing new sources of food, but often prevailing taste preferences are very hard to change and new products, particularly unconventional marine products, are not readily accepted, even when there is not much of anything else.

A final possibility to augment the ocean's food yield is by means of aquaculture, the production of aquatic organisms under controlled conditions. Dr. T. V. R. Pillay of the Food and Agriculture Organization estimated 1981 aquaculture production at some 9 million metric tons. Of this total, the culture of marine organisms, or mariculture, produces approximately one third.

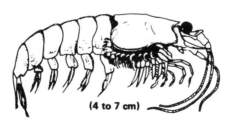

(4 to 7 cm)

Figure 2.6.   Krill (*Euphausia superba*).

Assessing the potential contribution of mariculture to marine protein supplies is a function of many factors. Progress is hampered by our limited knowledge of the sea and its inhabitants, but that is by no means the only deterrent. For another, the culture of marine organisms is usually an expensive operation: optimal conditions should be maintained, the animals need to be fed and kept disease-free, and their enclosures need to be cleaned. The process is labor as well as capital intensive (Fig. 2.7) and, as long as it remains so, investors will be interested in producing cash rather than protein crops. The bulk of world mariculture production for human consumption, as a consequence, consists of such species as

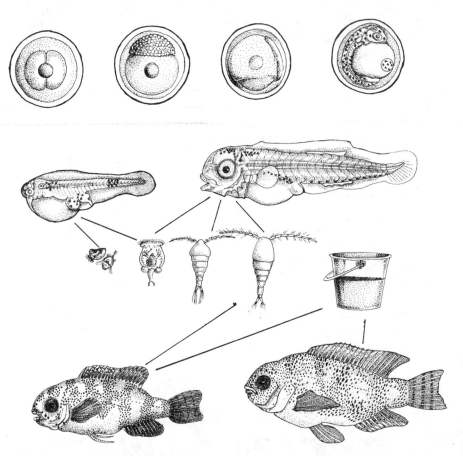

Figure 2.7.   Seafarming: the culture of the red sea bream (*Chrysophrys major*) in Japan. The continuous monitoring and feeding of larval and postlarval stages makes this an expensive venture. (Line drawing by Patricia Hansen.)

oysters and scallops, prawn and shrimp, and salmon—admittedly species that will not add much to protein shortages. In addition, there is no institutional framework for mariculture. It is a relatively new ocean use, in direct and mutually exclusive competition with longstanding ocean activities such as navigation, resource extraction, waste disposal, or even tourism. Its low priority in comparison to these traditional ocean uses is, in many instances, reflected in an unaccommodating legal regime, further deterring its development.

Despite these impediments, mariculture is already contributing a substantial and increasing share to the marine protein supplies of many countries, notably Japan and countries of Southeast Asia. It can provide significant supplies of marine protein on a global basis, provided that it is given the time, space, and resources needed for further development.

The exploitation of conventional and nonconventional marine species and the further development of mariculture could markedly raise the ocean's food potential. The Food and Agriculture Organization of the United Nations estimates that, by using the surplus potential, from 220 to 440 million metric tons of food could be harvested from the sea (Table 2.8), enough indeed to handle present and future needs. It is important to treat such estimates realistically, however, by taking account of the

Table 2.8. **Utilization and Potential of Fishery Resources (millions of tons)**

Resource	Food	Fishmeal	Discards	Postharvest Losses	Additional Potential
Marine					
Demersal	20	6	4–6		15
Pelagic	12	11			25–30
Other	3	—			2
Cephalopods (squid)	1	—			10–100+
Mesopelagic stocks	—	—			100
Krill	—	—			50–150+
Other crustaceans	2	—			1
Other molluscs	3	—			Very large
Freshwater species	11	—			5
Aquaculture	6	—			25–40+
Total	52	17	4–6	6+	240–450+

Source. Adapted from FAO, 1977.

time necessary for depleted stocks to rebuild as well as the difficulties associated with the exploitation of unconventional stocks. Both emphasize the need for drastic changes in current fishery management practices.

2.3. THE ECONOMICS OF FISHERIES

At one time in the development of European agriculture, each community had its commons, an area of grazing land open to all. As could be predicted, each farmer ran as much cattle as possible on the commons, because each additional animal accrued more wealth for him. Each farmer, in other words, pursued his own short-term interests, regardless of the long-term effects on him or his fellow farmers. The number of animals on the commons soon was much greater than the land could support. The economy of the farmer collapsed and the commons as a functioning, productive unit was impaired, if not destroyed. The commons ended with enclosure; grazing became restricted to a few.

The "tragedy of the commons" came about because land was a commodity: farmers had much more interest in rendering certain their own short-term welfare than in assuring the future productivity of the resource. Even though the commons is a concept in use many hundreds of years ago, many modern-day situations are comparable to it, and exhibit similar results—waste and misuse. Fisheries, unfortunately, are one of these.

Most fisheries, like the commons, are common property resources. Like the air we breathe or the water of streams, they are resources that can be used simultaneously by more than one individual. Moreover, no single user has exclusive rights to the resource or can prevent others from participating in its exploitation.

Another characteristic of fisheries is that they are renewable. In unexploited fisheries, natural mortality and recruitment are usually in balance, so that the population remains stable over the long run. Fishing represents additional mortality: to a certain extent fishermen replace natural mortality by taking fish that would otherwise have died from natural causes, but beyond that level, fishing reduces the population. The catch can be sustained indefinitely, however, if it, together with natural mortality, equals recruitment and growth. This sustainable yield depends on the size of the population and on the level of fishing effort: at low fishing levels population and natural mortality will be high, whereas higher levels of effort correspond to higher yields and a lower population. The point at which the highest sustainable yield is obtained is the maximum sustainable yield; levels of effort beyond this point will yield progressively lower catches because of the decreasing size of the population. As a result, in fisheries increased input (effort) does not necessarily increase output (fish), as is the case in most sectors of the economy. Instead, once beyond the maximum sustainable yield, increased inputs will yield lower output.

The common property (open access is the terminology preferred by economists) and renewability aspects combine to contribute to the mismanagement of fisheries. To show this, some exposure to elementary fishery economics is necessary. To keep the analysis simple, we will follow the descriptive model, perhaps not the best model available for management purposes but one that is easily understood and adequately enhances an understanding of the forces at work in the economy. Some additional simplifications need to be made: it is usually assumed that the price of fish remains constant, that the cost of effort increases linearly, and that the fishery consists of a number of identical vessels, which do not individually affect the catch. Such a fishery can be graphically represented as in Figure 2.8, where the sustainable yield curve doubles as the total revenue curve TR since the price of fish is held constant, and where TC shows the linearly increasing cost of effort.

When first exploited, a fish stock will yield a handsome profit to fishermen (the vertical distance between TR and TC). Since there is no limit on the number of fishermen that can enter the fishery—it is an open access fishery—this profit will entice more fishermen to participate, so that more effort is expended. The catch will increase but at one point the catch, together with natural mortality, will be larger than the sum of growth and recruitment, and the stock decreases in size. As the population declines, the cost of catching the remaining fish rises. Some fishermen may be forced to leave the fishery. Effort, as a consequence, is somewhat reduced, and the stock may begin to rebuild. This leads to lower costs, higher profits, and more fishermen, which again will reduce the stock. Eventually the fishery will arrive at an equilibrium (X) of population and effort along the sustainable yield curve. This point is an economic equilibrium, since total cost equals total revenue, and a biological equilibrium, since catch plus natural mortality equals growth and recruitment.

In open access fisheries, this equilibrium does not represent a preferable

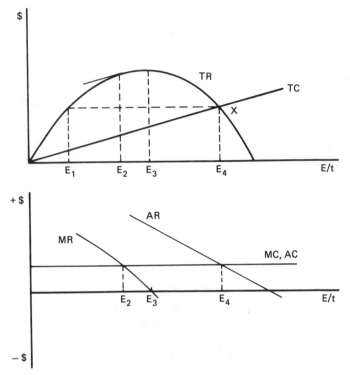

Figure 2.8.   The open access fishery. (Adapted from Anderson, 1977.)

situation. Indeed it denotes an amount of effort far larger than is strictly necessary. Note in Figure 2.8, for instance, that the yield of the open access equilibrium could also be obtained by applying E_1 units of effort, which, of course, would be much less expensive. Exploitation is pushed to the point where average costs are equal to total revenue. Profits are dissipated to pay for the unnecessary amount of effort. Even though representing an economic equilibrium, X therefore indicates a misallocation of resources.

In addition, the biological equilibrium is not desirable since it will usually occur beyond the maximum sustainable yield. The stock is theoretically not in danger at that point because mortality remains in balance with recruitment, but the population is low, the average age of the fish too small, in some instances recruitment may be affected, and the stock, in general, is not as able to withstand environmental or physical influences. Moreover, since the exploitation of a stock in practice does not nicely follow the path of sustainable yield, it is possible that a stock would become commercially extinct.

An examination of the behavior of individual fishermen under similar conditions reaffirms these observations. In the open access fishery, fishermen will produce effort as long as the return on it is greater than the cost of producing the last unit (return per unit of effort equals the product of price and average catch per unit of effort). The equilibrium average return, like the price in a purely competitive economy, is at the intersection of the market demand and supply curves. As seen in Figure 2.9, the supply curve of effort is the sum of the supply curves for individual vessels while the average revenue curve serves the same purpose as a demand curve. Under open access exploitation, the position of the supply and demand curves results in an expansion of effort larger than necessary, provided by too many vessels. Maximum economic yield, in contrast, would be obtained at the intersection of the industrywide supply curve and the marginal revenue curve. Stock externalities indeed are reflected in marginal revenue but not in average revenue, which is what the fisherman is concerned with. At maximum economic yield, the correct amount of fish is supplied by a smaller number of vessels. This generates a profit, unlike the open access fishery where it is dissipated among new entrants.

It can also be shown easily that the conditions for achieving optimum in welfare analysis are not met by the open access fishery. Social surplus is not maximized because the resources used by the fishery ought to be allocated to other segments of the economy. A fishery under sole ownership may result in an optimal allocation of resources but would not necessarily lead to social optimum. In fact, fishermen would probably lose in a switch from open access to sole ownership unless they were compensated through a redistribution of the profit of a sole owner.

The open access nature of fisheries and stock externalities cause overfishing and a misallocation of resources, but there are remedies. Sole ownership could be one, but there are very few instances where bringing a fishery under sole ownership would even be possible. A more practical way of eliminating the externalities is by means of regulations to obtain the correct amount of fish at the lowest cost. There are various institutional arrangements that attempt to meet this goal. Regulations affecting the size of fish such as mesh sizes, as well as closed seasons, closed areas, and quotas are favored by policymakers. Although these schemes may conserve the stocks, they will never lead to more efficiency.

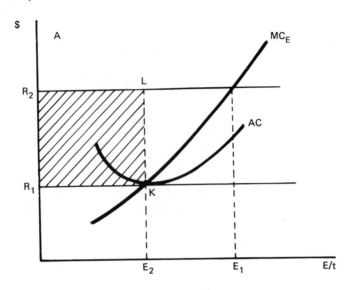

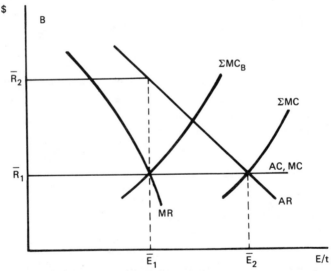

Figure 2.9.   The fishery and the individual fisherman. (Adapted from Anderson, 1977.)

In the case of quotas, for example, fishing seasons have been shortened substantially since fishermen, in an effort to obtain the largest possible share of the quota, are impelled to fish very intensively for a short period of time. Also favored by policymakers are regulated inefficiencies, which require fishermen to employ inefficient equipment to lower the total effort. As these methods permit a larger number of fishermen to remain in the fishery, some social welfare may be gained, but the use of inefficient equipment can, of course, never lead to an optimum allocation of resources.

Regulation by means of taxation is probably the best method to prevent the economic problems of fisheries. A tax on total effort, though difficult to implement, could shift the marginal cost curve up so that it would intersect the average yield curve at a preferable location (Fig. 2.9b).

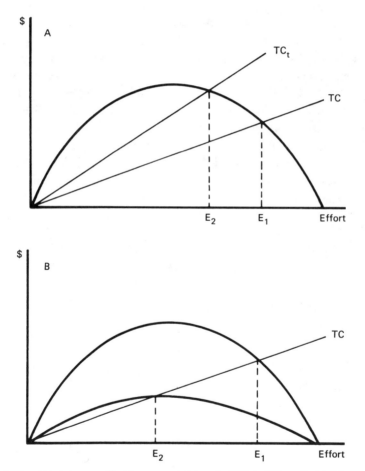

Figure 2.10. (A) A tax on effort increases total costs (TC) to TC_t so that total effort decreases from E_1 to E_2. (B) A tax of fish deflates the revenue curve (TR) so that total effort decreases from E_1 to E_2.

Similarly, a tax on fish could deflate the total revenue curve to obtain the same effect (Fig. 2.10). In both these cases, taxation would yield a rent to the managing authority. Of course neither method would be popular with fishermen. Limited entry programs such as licenses, effort shares, or fisherman's quota can also bring about a better allocation of resources, but these methods have the great disadvantage of shifting the burden of selection from market forces to the management authority.

The goal of regulation is to obtain the correct amount of fish at the lowest cost, and this invariably requires the reduction of fishing effort. Reducing effort can result in unemployment and the lay-up of fishing vessels, however, and these are not pleasant consequences for policy-makers to face. That is why many fisheries were allowed to be overfished until the stock became so small that fishing turned unprofitable, at which point fishermen would shift to other fisheries and aggravate the situation there. Once regulations became an absolute necessity, they included mostly schemes that were preferable from a political, rather than an economic, point of view. They treated symptoms rather than their cause, in many instances merely postponing inevitable depletion.

Even though very simplified, the preceding economic analysis provides some insight into the economic problems of fisheries. Fishery management schemes have largely ignored these problems or even magnified them through their single-minded pursuit of the largest physical yield and their use of politically acceptable regulations. Fishermen cannot be expected to willingly restrict their effort, for what they leave for tomorrow will be taken by others today. Even though self-destructive, this behavior is imposed by the open access nature of the resources. Unfortunately, this shortsightedness also applies on a much larger scale. Many nations indeed failed to curtail their fishing effort, realizing they would not be able to enjoy the product of their restraint. This philosophy, aggravated by a rising demand for fish, led to waste, misuse, and, in some instances, even conflict. In the case of the commons, the system was saved from total destruction by enclosure. It appears this may be the only way to guarantee a more optimal use of the ocean's food resources as well.

2.4. THE INTERNATIONAL LAW OF FISHERIES

Until the twentieth century, the international law of fisheries was based on the principle of the freedom of fishing. This made good sense: fish stocks were considered to be inexhaustible, so there was no need for anyone to regulate their exploitation. Yet there was not universal agreement on the issue, and the freedom of fishing principle was accepted only after long arguments. This was particularly the case in the North Sea, where conflicts between fishermen from different nations occurred as early as the Middle Ages. Usually these disagreements were solved by force rather than negotiation, preventing the development of an acceptable legal regime of sorts. As the North Sea became more intensively fished in subsequent years, the situation steadily deteriorated, until a disagreement over the right to fish for herring contributed in no small part to the outbreak of of the Anglo-Dutch Wars of the seventeenth century.

There was a concurrent disagreement on the issue of navigation, pitting two conflicting philosophies—national ownership and freedom of navigation—against each other. Since the Dutch did not want to pay for a license to trade and navigate or to fish, Holland's leading international lawyer, Hugo Grotius, spelled out the arguments in favor of freedom of the seas. "The sea, since it is as incapable of being seized as the air, cannot be attached to the possessions of any one particular nation," he wrote in *Mare Liberum*, published in 1609 when he was only 25 years old. This made good economic as well as philosophic sense and, since no nation could really enforce its claims over the sea, it was this doctrine that prevailed. It found gradual acceptance in the eighteenth and nineteenth centuries, remaining subject, however, to the restraint that each coastal state had sovereignty over an area of the sea adjacent to its coast, which became known as the territorial sea.

Whereas for 300 years freedom of the seas essentially meant freedom of fishing and navigation, the concept was broadened in the twentieth century to include the freedom to exploit and to pollute, and the freedom to do so irresponsibly. As a result, the freedom of the seas became an increasingly ambiguous concept. Traditionally, it had been based on what Hugo Grotius spelled out earlier: the waters of the sea are not susceptible to effective occupation, the resources of the sea are inexhaustible, and a specific use of the sea does not impair other uses; but these assumptions were no longer of unquestionable validity. Customary law, on which the international law of fisheries had been based for hundreds of years, could not evolve rapidly in response to emerging problems such as pollution and overfishing, and increasingly the law of fisheries became codified in multilateral and bilateral treaties, modifying the freedom of fishing principle by subjecting it to specific restrictions and conditions. Even so, disagreements on fishery issues remained common. Many coastal states, witnessing the decline of the fish stocks off their coasts, sought to ensure protection by means of unilateral measures that were considered directly in opposition to the principle of the freedom of fishing by the countries which saw their fishing operations drastically curtailed.

These and other topics were discussed at the 1958 Law of the Sea Conference in Geneva, which adopted two conventions that specifically address the legal regime of fisheries. On conservation issues, the key instrument was the 1958 Convention on Fishing and Conservation of the Living Resources of the High Seas. It defined conservation as "the aggregate of measures rendering possible the optimum sustainable yield from these resources so as to secure a maximum supply of food and other marine products." The treaty also emphasized negotiation and nondiscrimination in conservation matters, but beyond that remained relatively abstract since it treated fisheries in general rather than particular stocks in specific areas. Even so, it provided an impetus for the conclusion of additional conservation agreements, whether on a regional basis or treating a particular species on a global basis.

In terms of allocation, the Geneva Conventions remained silent, aside from stating, in the 1958 Convention on the High Seas, that the freedom of the high seas included the freedom to fish. Other than that, there never was any agreement on the subject. Some coastal states, eager to protect their fisheries from depletion, favored extensive coastal jurisdiction; others, particularly the distant-water fishing nations, wanted to prevent any extension that would jeopardize their operations. As a result, some nations claimed 200-mile zones, others 12 miles, and the majority 3 miles. A second attempt to settle the question was undertaken at the second Law of the Sea Conference in Geneva in 1960. This time a proposal entailing a 6-mile territorial sea and an adjacent 6-mile fishing zone came very close to being approved, but finally failed by a single vote. It was the consequent void in allocation matters in the law of fisheries which led to a new generation of fishery conflicts, ranging from cod wars off the Icelandic coast to a tuna war of sorts off Latin America.

The Geneva Conventions set some general guidelines on the law of fisheries but certainly did not establish a comprehensive system. In fact, it is doubtful whether this would even have been possible, since the international law of fisheries, like any other system of law, is not a self-serving and self-contained entity. As an instrument of social order, it functions within a context of natural, technological, economic, and political factors. The Geneva Conventions hardly took account of these aspects and, as a result, remained far from sufficient in promoting the orderly conduct of fishing.

A partial addition was provided, however, by regional agreements, some of which were concluded along the guidelines of the Geneva Conventions, others of which had been in existence before the onset of the 1958 Law of the Sea negotiations. Such a regional approach made sense because it allowed for more flexibility, needed because fish stocks are mobile and do not respect artificial boundaries. Furthermore, fishing patterns and techniques differ widely from one region to another, as do a variety of social and political factors.

Conservation agreements were established in all of the world's oceans (Table 2.9). Most of them set up functional commissions, designed to ensure the conservation of the resources. But it didn't quite work out that way. The fishery commissions differed greatly from one another in their terms of reference, their powers, and their structure, but in common nearly all failed in their missions. In the field of fisheries, as in other areas of ocean law, states were extremely reluctant to give up any of their prerogatives in favor of international organizations. As a consequence, most commissions did not get the powers needed to succeed in their tasks, and the fish stocks they sought to protect were depleted.

Realizing that the international conservation regime was unable to halt overfishing, many coastal states decided to further extend their fishery jurisdiction in the early 1970s, this time to 50 and 200 miles, far beyond the generally accepted 12-mile limit. Again, bitter conflict between coastal states and distant-fishing nations ensued, but the latter were fighting a losing battle. In a remarkably short time, the international community turned overwhelmingly in favor of extended jurisdiction, a change which drastically altered the allocation and conservation patterns in fisheries.

There are no provisions for this change in the current international law of fisheries. The issue was debated at the recently concluded third Law of the Sea Conference (1982), and since it received consensus there, it will become international law when the new convention goes into effect. From a legal point of view, fisheries now are in a transitional stage, marked by a trend from nonterritorial to territorial restrictions. Hundreds of years ago, this would have made no sense. Today, it may be the only viable way to ensure a more intelligent use of the sea's food resources.

Table 2.9.   **International and Regional Fishery Commissions**

BSSC	Baltic Sea Salmon Committee
CARPAS	Regional Fisheries Advisory Committee for the Southwest Atlantic
CECAF	Fishery Committee for the Eastern Central Atlantic
GFCM	General Fisheries Council for the Mediterranean
IATTC	Inter-American Tropical Tuna Commission
IBSFC	International Baltic Sea Fishery Commission
ICCAT	International Commission for the Conservation of the Atlantic Tuna
ICSEAF	International Commission for Southwest Atlantic Fisheries
INPFC	International North Pacific Fisheries Commission
IOFC	Indian Ocean Fisheries Commission
IPFC	Indo-Pacific Fisheries Council
IPHC	International Pacific Halibut Commission
IPSFC	International Pacific Salmon Fisheries Commission
IWC	International Whaling Commission
JKFC	Japan–Republic of Korea Joint Fisheries Commission
JSFC	Japan–Soviet Northwest Pacific Fisheries Commission
MCBFC	Mixed Commission for Black Sea Fisheries
NAFO	Northwest Atlantic Fisheries Organization
NEAFC	North East Atlantic Fisheries Commission
NPFSC	North Pacific Fur Seal Commission
PCSF	Permanent Commission of the Conference on the Use and Conservation of the Marine Resources of the South Pacific
WECAFC	Western Central Atlantic Fisheries Commission

With 200-mile fishing zones in effect, approximately 90 percent of all fisheries will come under national jurisdiction. Fish stocks will no longer be common property resources, open for exploitation to all. The legal regime is adapting to the changes necessitated by their common property nature and is evolving from a system of open access to enclosure. This will not necessarily result in optimal exploitation or equitable allocation, but at least there is somewhat more of a guarantee that the ocean's food resources are not wasted and misused. In a world that is made up of sovereign states, a system that would best serve the interests and promote the welfare of the world at large unfortunately remains unrealistic.

Chapter Three

Minerals from the Sea

During the last few decades, considerable interest has developed in the mineral wealth of the ocean. It has become obvious, for instance, that present industrial growth cannot be met by land-based resources alone, and we are increasingly turning to the sea as a possible source of additional materials. This has resulted in the intensive exploration for and development of a number of marine minerals, a process made possible by the awesome growth in science and technology.

Only recently has an appreciation grown for the full potential of the sea's mineral resources, with estimates invariably ranging in the billions or even trillions of tons. Yet despite these optimistic figures, many of the sea's minerals are not economically recoverable, though there is little doubt that once needed, these supplies can and will be recovered. For the moment, however, exploitation has been confined principally to continental shelf areas.

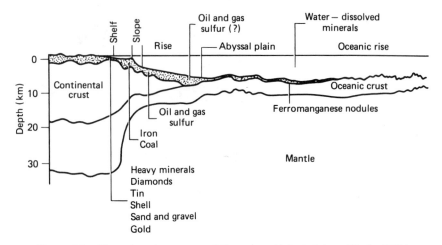

Figure 3.1.   The mineral resources of the ocean. (Adapted from Wenk, 1969.)

In addition to oil and gas, which of course top the marine minerals list in terms of value, minerals from the sea include those extracted from seawater, surface deposits, and consolidated subsurface deposits. A short review of the extent of these resources is presented in Section 3.1 (Fig. 3.1).

3.1.   THE RESOURCES
Seawater

Because the oceans are so wide and so deep, statistics on their total resource potential are very impressive. Each cubic mile of seawater, for instance, contains about 165 million tons of solids, making the 350 million cubic miles of seawater the world's largest continuous ore body.

All chemical elements are present in seawater, though only a few of the major constituents lend themselves to commercial extraction. Since sodium and chlorine account for 85 percent of the sea's dissolved salts, it is not surprising that sodium chloride, or common salt, was the first mineral removed from the sea, undoubtedly since prehistoric times. Magnesium, the third most abundant major element in the oceans, has been extracted since the late nineteenth century. It is taken out of seawater in an alkaline condition and removed by precipitation. A substantial amount of bromine—used in antiknock compounds for gasoline—is also obtained from seawater, even though this element is present in very small concentrations. The availability of other sources of bromium coupled with the reduced demand for leaded gasoline may phase out this industry, however.

Most other chemical elements remain beyond our ability to extract from seawater commercially, though uranium removal may become a possibility in the future by means of a new Japanese technique. The value of other minerals such as gold, silver, copper, and zinc in 1 cubic mile of seawater is substantial, but an extraction plant handling that amount of water would need to process more than 2 million gallons per minute every minute of the year, which would make its operating costs far above the combined value of all its products.

Of all the sea's resources, the most priceless is water itself. It is expensive to extract fresh water from the sea, but with water requirements for domestic, agricultural, and industrial purposes constantly on the rise, desalinization becomes increasingly attractive. As of 1981, the world-wide desalinization capacity of all plants producing more than 25,000 gallons per day amounted to 2 billion gallons of fresh water per day. Most of these desalinization plants are in operation in water-deficient areas such as the Middle East.

Surface Deposits

The mineral resources of the seabed, unlike those of the essentially uniform overlying water, occur primarily in scattered, highly localized deposits. One exception, however, is found in the abundant sand and gravel resources of the continental shelf. Sand and gravel are used extensively in construction, and with land reserves becoming limited in some areas, it was quite natural that the construction industry moved offshore to secure its needs. This has been the case particularly in western Europe and Japan.

The extent of sand and gravel resources has been tentatively estimated for the United States and the world, both onshore and offshore, and compared with projected demands. Table 3.1 shows that, according to these estimates, economical land sources of sand and gravel will be depleted by the end of the century, but it is clear that the exploitation of offshore deposits could substantially increase the resource base. There is no way to substantiate these impressive numbers, mainly because the resources have been mapped in only a few areas such as the northeastern United States and western Europe.

Table 3.1. **Comparison between (apparent) Marine Sand and Gravel Deposits and (apparent) Land-Based Sand and Gravel Deposits**

Variable	U.S.	World
Annual demand (1970)	1.02	7.45
Cumulative demand to 2000	76.70	535.00
Apparent land resources	67.00	333.00
Apparent marine resources	1,690.00	31,000.00

Source. Cruickshank and Hess, 1976.
Note. The units represent gross market values in billions of 1972 dollars. They do not indicate economic reserves; instead, they are used for comparative purposes.

Marine sand and gravel deposits traditionally are mined by dredging, which is an attractive technique because of its mobility. But this type of exploitation tends to interfere with other ocean activities and, not surprisingly, disturbs the marine environment. The potential impact on marine life is very difficult to assess and remains a poorly known area of environmental concern. Current investigations indicate that the adverse effects may be more damaging to the physical features of the coastal zone than to its ecology, but there is not enough evidence to substantiate this claim. It is well known, however, that the damaging effects of offshore sand and gravel exploitation can be minimized if the operations are supported by adequate information about the deposits and take account of the environmental conditions in the exploitation area, but this is certainly not always an established practice.

Detrital minerals, also found on the ocean bottom, are those heavy minerals of economic value, separated from their lighter gangue by high-energy wave action in shallow waters. None of these deposits is a true ocean mineral since they are found in drowned shoreline extensions, which were submerged by rising seas after the last Ice Age. A number of these deposits are currently commercially exploited.

Iron sands, rich in such minerals as titanomagnetite, ilmenite, rutile, zircon, and monazite, occur in beach deposits in many areas, particularly India, Egypt, Brazil, Australia, New Zealand, and a number of East Asian countries. The Australian deposits constitute the world's principal source of rutile, a titanium mineral used to produce white pigments, as well as zircon and monazite. Iron sands also occur in large quantities as beach deposits off the west coast of New Zealand, where they are mined for export to Japan.

Cassiterite, an important tin ore, is a residual mineral of the weathering of granites. It has been mined offshore in Southeast Asia since the start of this century, although only in the past few years have operations ventured beyond sheltered bays. Until recently, diamonds were mined off the Namibian coast. They are found in offshore deposits at depths of up to 40 m, but their mining was never entirely profitable in view of the extreme sea conditions and relatively low offshore concentrations.

Subsurface Deposits

Marine subsurface minerals include fluids and soluble minerals that can be extracted through boreholes as well as consolidated deposits such as coal, iron ore, and other minerals found in veins.

Of all marine minerals, oil and gas are the most important in terms of value. The history of marine drilling for oil goes back to 1938 when the Creole field was discovered in the Gulf of Mexico, about 1 mile from the coast of Louisiana, but it was after the Second World War that offshore hydrocarbon exploitation got its real impetus. Today, the oil industry has expanded to deep and shallow waters all over the world.

Several criteria need to be satisfied before petroleum can accumulate in recoverable quantities. First, there must be a rich source of organic material, preserved until buried by sediments. Second, this organic matter must be subjected to temperature and pressure increases, necessary to convert it to liquid or gaseous hydrocarbons. These must then move from the fine-grained source beds to coarse-grained, permeable reservoirs, and a trap needs to be present to prevent them from leaking out. Finally, the sediments should not be altered or metamorphosed by folding or geothermal heating, since this would destroy the oil or permit its escape through fractures.

Many offshore areas appear to satisfy these requirements since oil largely is derived from diatoms, whose remains became entrapped in marine sediments. Moreover, as a result of the vicissitudes of geologic history, oil tends to become lost with time so that sediments less than 150 million years old, which predominate offshore, offer the most promising oil reservoirs.

Present drilling technology and the price of fossil fuels limits the commercial recovery of offshore oil to the continental shelf, but it appears that vast reserves may be located beyond. The great prisms that lie at the bases of nearly all continental slopes, for instance, are probably a great untapped source of oil. All conditions for oil accumulation indeed seem to be right: the sediments, commonly a few miles thick, are young and rich in organic matter and contain numerous lenslike sand bodies that would make fine reservoirs and traps.

The oil potential of the deep sea, on the other hand, is not well known. The vast red clay-covered expanses of the Pacific Basin are presumed to contain little oil. Red clay is poor in organic matter since most of the sediments are produced beneath waters of low productivity. The diatom oozes located beneath the Antarctic Ocean, in contrast, may contain a lot of oil, and it also appears that the Atlantic Basin holds oil reservoirs. The technological difficulties associated with recovery of these reserves are enormous, but once the price is right, such problems undoubtedly will be solved.

Given the many uncertainties about the oil potential of these areas, there is no global estimate on the magnitude of offshore hydrocarbon

Table 3.2.   **Estimated World Production of Crude Oil and Natural Gas**

Year	Onshore Production	Offshore Production	Total Production	Offshore as a Percentage of Total Production
Petroleum (thousands of barrels)				
1970	13,969,418	2,749,290	16,718,708	16.4
1972	15,360,100	3,240,645	18,600,745	17.4
1974	17,132,097	3,405,630	20,537,727	16.6
1976	17,666,395	3,525,145	21,191,540	16.6
1978	17,963,201	4,195,050	22,158,251	18.9
1980	16,763,166	5,006,880	21,770,046	23.0
1982*	14,520,430	4,876,400	19,396,830	25.1
Natural Gas (millions of cubic feet)				
1970	32,832,942	5,261,019	38,093,961	13.8
1972	39,506,581	4,017,978	43,524,559	9.2
1974	40,934,670	6,243,850	47,178,520	13.2
1976	37,475,038	10,830,646	48,305,684	22.4
1978	42,240,355	9,508,980	51,749,302	18.4

Source.   American Petroleum Institute, 1982.
* Preliminary estimates.

deposits. This type of information is available only for intensively explored regions such as the Gulf of Mexico and the North Sea. Despite this lack of quantitative information, the ocean most likely contains very substantial amounts of oil and gas. One recent estimate indicated that offshore areas hold about 23 percent of the world's crude oil reserves and 14 percent of the natural gas reserves but, since offshore exploration and exploitation have been restricted principally to relatively shallow waters, this could turn out to be an underestimate.

The contribution from offshore oil wells to world oil production in 1982 approached 25 percent, while the offshore gas share experienced a sharp decline (Table 3.2) This production represents more than 90 percent in value of all minerals recovered from the oceans.

Other subsurface deposits mined offshore include sulfur and potash. Sulfur, one of the most important industrial chemicals, is found in the cap rock of salt domes buried within continental shelf and seafloor sediments. It can be recovered inexpensively by melting with superheated water and then forcing it up with compressed air.

Consolidated minerals such as coal, iron ore, nickel, copper, and tin are mined from shafts driven as far seaward as 30 miles from the British and Japanese coasts. In the future, shafts may be driven directly from the seabed to recover ore deposits located far from land.

Nodules

Of the minerals of the deep sea, none has received quite as much attention as ferromanganese nodules. These dark, potato-sized lumps were discovered more than a hundred years ago by scientists during the *Challenger* expedition. At the time, nobody attached particular attention to them, and they ended up being classified as mineral curiosities in the British Museum.

Ferromanganese nodules were reassessed during the late 1950s when, after analysis, it became clear that they contained concentrations of a number of metals, comparable to the levels of terrestrial deposits. In addition, deep sea explorations revealed that nodules were found in every major ocean, sometimes in concentrations from 50 to 100 kg/m², and thus formed one of the largest mineral deposits on Earth. Even though current knowledge on the total amounts of ferromanganese nodules is based on sketchy data, estimates range in the trillion-ton range for the Pacific Basin, and perhaps twice as much worldwide.

Quite naturally, these figures fascinated scientists, business executives, and politicians alike. Scientists were interested primarily in finding out how nodules obtained their high metal concentrations, typically averaging 20 percent of manganese, 6 to 10 percent of iron, 1 percent of copper, 1.2 percent of nickel, and 0.4 percent of cobalt. The process is still not entirely understood. Several theories have been advanced postulating that the metal concentrations are derived through adsorption from metal-rich sediments, through the adsorption of metal cations by negatively charged manganese oxides, through direct concentration from seawater, or perhaps through microbiological activity. Metal concentrations in nodules vary from one region to another, but again it is not known why this occurs. If scientists were able to answer this and a number of other questions, they would perhaps be able to pinpoint where the richest deposits are located, something that would, of course, be greatly appreciated by the industry.

Despite many uncertainties, commercial interest in nodules was large from the onset since four metals—manganese, copper, nickel, and cobalt—appeared to be present in commercial quantities. None of these metals is really in danger of being depleted on land, but in most instances land deposits are located in Third World countries, and nodule mining offered industrial nations the prospect of reducing their dependence on Third World supplies. Before that could happen, however, an entirely new technology for mining and processing needed to be developed since the nodules were located at depths averaging 4 to 6 km—far beyond the reach of conventional technology (Fig. 3.2). The trend for the firms interested in deep ocean mining has therefore been to form international consortia, which permit risks, capabilities, and investments to be shared while drawing broadened government support in the absence of international agreements.

The richest nodule deposits have been found in the Central Pacific Ocean, about 1500 km southwest of Hawaii. Most of the industrial consortia have concentrated their efforts in this area, characterized by depths ranging between 5 and 6 km and a sea bed covered by nodules rich in copper and nickel. The largest amount of nodules recovered from this area totals 885 tons, a far cry from the 3 million metric tons per year required for commercial recovery. Timetables for ocean mining are slipping, not in a small part as a result of legal uncertainties, and it now appears there will be no commercial recovery until the 1990s or perhaps the end of the century.

In addition to ferromanganese nodules, phosphorite nodules may become a marine mineral of commercial importance. Phosphorite nodules

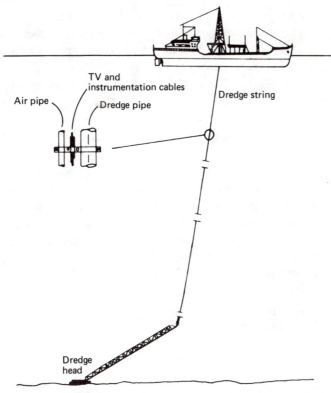

Figure 3.2.   A hydraulic system proposed for mining manganese nodules. Air is forced into the dredge pipe. As it rises to the surface, it expands, bringing the nodules to the surface.

are found in many continental shelf areas, particularly off the coasts of California, Peru, Morocco, and New Zealand. Data on their distribution are limited to these areas. For the southern California region, for example, reserves have been estimated at 11.2 million tons of phosphorite and 45 million tons of phosphate sands. On the Chatham Rise off New Zealand, preliminary estimates suggest that well over 100 million metric tons of phosphorite are available.

Despite the abundance of phosphorite deposits close to shore, where many are exploitable with modern dredging techniques, they have not been developed, particularly since their grade is rather low and the impurities are difficult to remove. In addition, there are still many high-grade deposits on land. In the United States, where annual production averages 100 million tons, land-based phosphate reserves amount to

14.5 billion tons, while reserves in northern Africa may be four times as large. There are special circumstances, however, which may make marine phosphorite exploitation attractive. This appears to be the case in New Zealand, which currently imports 1 million tons of phosphate annually and which has the world's highest per capita phosphate consumption.

Metalliferous Sediments

A particularly interesting example of metalliferous sediments of potential commercial significance are those associated with hot brines of the Red Sea, discovered in 1963 by oceanographers of the British research vessel *Discovery II*. At depths of 2000 m they found localized pockets of brine, with a salinity 10 times that of seawater and temperatures over 60°C. The underlying sediments were found to be black, amorphous iron oxides, rich in other metals such as zinc, lead, and silver. The sediments under one of these brine pools, the Atlantis II deep, contain an average of 28 percent iron, 2.5 percent zinc, 0.5 percent copper, 0.1 percent lead, and 90 ppm (parts per million) silver. The value of the sediments in the upper 10 m of this deep has been estimated to exceed several billion dollars.

Hot brines have a complex geological history. In the case of the Red Sea, they were formed millions of years ago when warm waters from the interior of the Earth, as a result of seafloor spreading, penetrated into a series of depressions along its axis. The water passed through thick salt deposits and volcanic material, where they took salt in solution as well as a number of metals that formed complexes with chlorine. As a result of their high salinity, they remained stratified from the overlying waters, forming brine pools. At least one mining company has undertaken feasibility tests in this area.

More recently, metal-rich muds have also been found to be common on mid-oceanic ridge crests, where they rest on top of the cooled lava. The deposits contain principally iron and manganese, which are not of much value, as well as a number of metals such as copper, nickel, and zinc in smaller concentrations. These minor components are scarce on land, so that the mid-oceanic ridge muds may also become a usable mineral resource competitive with ore bodies on land.

The extent of these deposits has not been determined, but initial investigations reveal that they are very extensive, having been discovered along the East Pacific Rise, the Galapagos Rise, and the Juan de Fuca Rift. Moreover, hydrothermal fluids rise through fissures along mid-oceanic rifts continuously, so that a number of "nonrenewable" resources are in fact constantly being renewed, albeit at a very slow pace. Commercial mining of these deposits is still decades away, though their investigation may give clues as to where similar deposits can be found on land.

3.2. THE REGULATION OF MARINE MINERAL DEVELOPMENT

The management of marine minerals differs from fisheries in that most minerals are nonrenewable and static. The nonrenewability aspect implies that the resources are subject to possible depletion: what is taken one year reduces what is left for the following years, so that ultimately the resource can be exhausted. Decisions need to be made to prevent this from occurring too soon, or perhaps too late. The static nature of most marine minerals is also important because it allows them to be allocated unambiguously, unlike fish stocks, which move freely across artificially imposed boundaries.

Until some forty years ago, nobody had shown much interest in the mineral wealth of the ocean floor, simply because most mineral needs could adequately be supplied by the land. But this situation changed rapidly after World War II when the United States asserted jurisdiction over the mineral resources of the continental shelf. To secure these minerals, and primarily new sources of petroleum, President Truman in 1945 issued two proclamations, one of which claimed the exclusive right of the United States to exploit the resources of the sea bed and the subsoil of its continental shelf. The other proclamation concerned the resources of the overlying waters, particularly fisheries, over which no exclusive rights were asserted.

These proclamations led to considerable controversy. Some nations reacted with suspicion, others with indifference, and a number of them saw in it an opportunity to adjust supposed injustices. This was particularly the case with some Latin American countries which possessed hardly any continental shelf but were still interested in protecting their coastal resources. Hence a few years after the Truman Proclamation, Chile, Peru, and Ecuador signed an agreement to extend their sovereignty over the sea and the sea bottom out to a distance of 200 nautical miles, claiming, in effect, a territorial sea of that size. Even though the United States and most other countries never recognized these claims, the policy set by the original Truman Proclamation was followed by many other states, who thus recognized that coastal jurisdiction over the resources of the continental shelf was a fair and reasonable thing to assert. The practice was subsequently codified in the 1958 Geneva Convention on the Continental Shelf, which gave coastal states sovereign rights over the continental shelf for exploration and exploitation purposes.

Less clear, however, was the definition of the continental shelf, or how far it extended. The Truman Proclamation did not define this but a statement released by the White House on the same day explained that "generally, submerged land which is contiguous to the continent and which is covered by no more that 100 fathoms of water is considered as the continental shelf." The Geneva Convention was even more ambiguous, describing the continental shelf as

> the seabed and subsoil of the submarine areas adjacent to the coast but outside the area of the territorial sea to a depth of 200 m, or beyond that limit, to where the depth of the superadjacent waters admits of the exploitation of the natural areas of said areas, [and] to the seabed and subsoil of similar submarine areas adjacent to the coasts of islands.

The Convention, in other words, did not specify where the continental shelf was supposed to end. At the time this was not considered a serious flaw because no one foresaw mining much beyond the 200-m depth. With present technology, however, almost the entire sea bottom could be interpreted to fall under this definition, which certainly was not the intention of the nations that drafted the rules.

The 1958 Convention also addressed the issue of dividing the continental shelf between adjacent and opposite states, but again, did so in a fairly ambiguous manner. It declared that such divisions should preferably be made by agreement. If this proved impossible, an equidistant line was proposed, unless special circumstances justified some other solution. The Convention defined what it meant by equidistance but did not explain what special circumstances entailed, leaving the division procedure open to a variety of interpretations.

In 1958 this system was probably adequate but, when it became apparent that the continental shelves contained resources far beyond what had been anticipated, delineation grew in importance. States with concave coastlines, for instance, realized that equidistance would yield them a proportionally small part of the continental shelf and its resources. As a result, the delineation rules proposed by the Geneva Convention were often challenged, most notably in the *North Sea Continental Shelf* cases between the Netherlands, Denmark, and Germany, a controversy which reached the International Court of Justice in 1969 (Fig. 3.3). In its decision, the Court agreed with Germany that equidistance was not an absolute rule and that other factors, such as the configuration of the coastline, the structure of the continental shelf, and to some degree proportion should be taken in account by states dividing up their continental shelf areas.

There are some 300 potential territorial sea or continental shelf boundaries, more than half of which divide the continental shelf of opposite states. Less than 25 percent of these boundaries have been negotiated; the rest are either in dispute, in some state of negotiation, or not being discussed. Most continental shelves, in other words, still need to be delineated. The principles that will be used in these instances are not certain, though it can be assumed that coastal states will support the principle which yields them the largest possible part of the pie. Perhaps this type of philosophy is best served by ambiguous rules, but with technology now permitting marine mineral development far beyond traditional depths, clearer definitions will be needed. Unfortunately, the new Law of the Sea Convention does not appear to make much progress in this respect.

The legal status of the deep sea bottom is also uncertain or, more accurately, became so when the potential of deep sea mining attracted more attention. Well aware of the imminent scramble for deep sea minerals, Malta's ambassador to the United Nations suggested in 1967 that the United Nations declare the deep sea bottom to be the "common heritage of mankind," an area not subject to appropriation by any nation for its sole use. The international community agreed, expressing its consent in a General Assembly Resolution, thereby changing the legal status of the deep sea from a legal no-man's land (*res nulius*) to a common area (*res communis*).

This agreement did not solve the problems for very long, however. The resolution was not legally binding, and many nations gave it different interpretations. Third World countries, for instance, clung to the tenet that the riches of the deep sea are the common heritage of mankind, demanding that all activities to exploit them be supervised by an international agency. The industrial nations, on the other hand, rejected this internationalization of the deep sea, arguing that they would have to come up with the skills, technology, and capital to mine nodules while the developing nations, dominating the agency, would reap the benefits.

These conflicting views were the major obstacle in the Third Law of the Sea Conference, recently concluded after nearly a decade of arduous negotiations. The convention adopted in 1982 provides for the creation of an International Sea-Bed Authority and an operating branch, called the Enterprise, as insisted upon by Third World countries. Deep sea development will take place under a parallel system; a company will submit two similar mining sites to the Authority, which will chose one

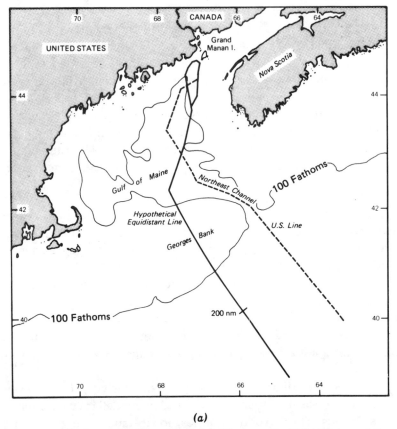

(a)

Figure 3.3.   (*a*) The division of the continental shelf between the United States and Canada as perceived by the U.S. government (dashed line) and the Canadian government (solid line).

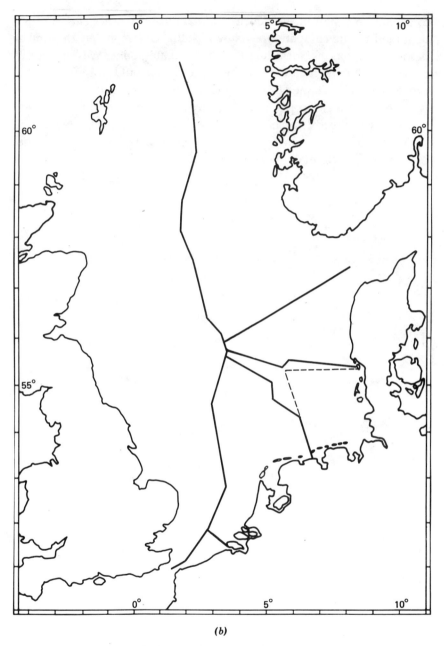

(b)

Figure 3.3. *(continued).*   *(b)* The division of the North Sea continental shelf by the Netherlands and Denmark (dashed lines) and the current division of the North Sea continental shelf among its riparian states (solid line).

to exploit on behalf of the developing nations. The other site can then be mined by the company for its own profit. Unfortunately, a number of industrial nations, including the United States, some West European nations, and most members of the Soviet bloc, could not accept these provisions and abstained or voted against the treaty. Until a compromise is reached (if ever), deep sea mining may well be regulated by unilateral mining laws, already implemented by such nations as the United States, Germany, and Japan. However, as observed earlier, the first commercial operations are still many years in the future.

Chapter Four

Waste Disposal in the Sea

F rom earliest times, the waters of the Earth have been a natural place to discard unwanted wastes. Until recently, this created few problems because natural waters have the capability of purification, but the serious condition of many streams, rivers, and lakes now is universally recognized. Even the seas are threatened by pollution from the great centers of population and industry along the shores as well as the many vessels crowding them.

The extent and effects of this pollution are imperfectly known. In some instances, biological damage has been demonstrated clearly, but in general far too little is known about the behavior and effects of marine pollutants to make any definite conclusions. Our knowledge of marine pollution is limited to some relatively narrow and ambiguous assessments, at times indicating some degree of danger yet often short of being sufficiently conclusive to prompt effective action.

It is even difficult to define marine pollution. Exactly what is a pollutant? Can a pollutant harmful in one situation be beneficial in another? A United Nations report defined marine pollution as "the introduction by man, directly or indirectly, of substances or energy into the marine environment (including estuaries) resulting in such deleterious effects as harm to living resources, hazards to human health, hindrance to marine activities, including fishing, impairment of quality for use of sea water and reduction of amenities." Confusing as it sounds, this definition clearly links pollution with harm and with man, and states the principal areas in which harmful effects may be experienced.

Pollution enters the ocean in four general ways. The first is by means of vessels which, in the course of normal operations, discharge a variety of pollutants, most significantly oil. In addition, vessels pose a threat of accidents such as collisions and groundings. Although they contribute a relatively minor share to ocean pollution, accidents cause the most visible forms of marine pollution, and therefore have received a disproportionate share of public attention. The second way pollutants reach the ocean is by deliberate discharges of harmful substances from vessels or platforms, a practice generally referred to as dumping. The third route includes sewer and industrial outfall, rivers, and land runoff. Material can be discharged into a river many miles upstream and be carried to the ocean by the river. These pathways are collectively grouped as land-based sources. Finally, significant amounts of pollutants reach the ocean through the atmosphere, most of them by rainfall. Of all routes this is the most difficult to trace to its source and to monitor with any degree of precision.

Marine pollutants can be classified into two important categories: degradable and nondegradable. Substances in the first group can be broken down in the marine environment and do not form any serious problem if they are discharged in reasonable quantities. Widespread degradable discharges include domestic sewage and effluents from pulp and paper mills or food processing plants. Nondegradable pollutants, in contrast, cannot be broken down and thus persist in the marine environment for a long time. Among this group are many pesticides, heavy metals, and numerous other industrial products. The many types of marine pollutants and their effects on the sea, as far as ascertainable, are briefly reviewed in the following section.

4.1. TYPES OF MARINE POLLUTION

Domestic Waste

Domestic waste is a turbid liquid with a content of particulated, finely dispersed colloidal and dissolved material from three main sources: kitchens, laundry rooms, and bathrooms. Wastewater from kitchens contains carbohydrates, fats, and proteins along with small amounts of other wastes. Waste from laundry rooms consists of small amounts of sand, dust, traces of oil, fat, textile fibers, and, most important, different kinds of soaps and detergents. Finally, domestic waste also includes effluents from sanitary equipment where excretion products are released into the sewer.

Before discharge, these wastes can be treated to protect inland water resources. Waste treatment may include removal of the sludge and suspended particles, the oxidation of polluted organic material, sterilization of the effluent, and removal of nutrients. In most instances, only the first one or two steps are taken, in which case large amounts of sludge are generated. Most communities near the ocean discharge their sewage into the sea without any treatment. It is impossible to estimate how much of this waste enters the ocean, but the fact that domestic waste flows in developed countries range up to several hundred liters per capita per day gives some indication of the enormous amounts involved. The effects of domestic waste disposal on the marine environment depend on the physical, chemical, and biological conditions of both the seawater and the waste. Some generalizations can be made, however. The first effects of sewage disposal into seawater are mainly of a physical and chemical nature. A variety of chemical reactions such as coagulation, precipitation, and flocculation take place. The effects of these processes can be harmful but they remain more or less local, concentrated in the vicinity of the outfall.

More serious are the primary and secondary effects of the degradation of the large amounts of organic material discharged in domestic waste. The primary effect of biodegradation is a reduction of the available oxygen in the water. In restricted waters, where there is not enough mixing, a full depletion of dissolved oxygen may occur. Biological oxidation processes continue beyond this level, first at the expense of nitrates and sulfates, which are thereby reduced. Poisonous hydrogen sulfide may then appear, preventing the existence of all higher life forms in the water. Such drastic effects are not observed in the ocean, but reduced oxygen levels may influence coastal fisheries since most commercial fish will remain only in water with a sufficiently high oxygen content.

The second effect of biodegradation is caused by the increased amounts of nutrients that remain after the breakdown of organic materials. The presence of excessive amounts of nutrients may indeed lead to eutrophication. In this process, the high fertility of the water permits rapid and excessive blooming of phytoplankton. Initially oxygen is overproduced, but at the termination of the bloom the decomposition of the algae reduces dissolved oxygen levels, causing mortality among fish and organisms that cannot survive under these conditions. In addition, the turbidity caused by the plants may increase to such a degree that virtually no light penetrates and photosynthesis is diminished. The eutrophication process will have the most drastic effects in restricted waters where nutrients can build up, but even in areas where dispersal is possible the effects on fisheries can be discernible. From a successional point of view, for instance, the niches of a community nearest the outfall tend to be filled with pioneer species: animals that adapt more easily to changes in the environment such as lower oxygen levels or increased nutrient concentrations. These animals tend to be small, short-lived, and low-valued. As a result, the yields from enriched areas, though possibly greater in terms of weight, are lower in terms of value. Eutrophication has been observed in estuaries and occasionally along the coast, but it is unlikely that it would ever extend into the open ocean.

Untreated wastes contain large numbers of bacteria and viruses which enter the marine environment upon discharge. The degree of bacterial pollution caused thereby is generally related to the number of coliforms. These bacteria are not dangerous to human health but their number indicates the presence of microorganisms that can cause diseases of various kinds. Such organisms can be concentrated by filter feeders such as molluscs and are transferred to humans when raw shellfish are consumed.

Finally, primary treatment of domestic waste generates large amounts of sludge, and the disposal of sludge in the ocean can create problems as well. Not only will sludge disposal reduce oxygen levels and possibly photosynthesis, but its settling may modify the structure of the sea bottom and may cause anaerobic conditions in or near the bottom. In such instances the composition of benthic communities will be changed and usually impoverished.

Industrial Waste

The inventory of industrial pollutants is far from complete. This should come as no surprise in view of the enormous amounts of waste products discharged by industrial sources, many of which ultimately reach the oceans.

Industrial waste products can be classified as organic and inorganic pollutants. Both groups are reviewed in this section. Two important classes of organic pollutants, hydrocarbons and chlorinated hydrocarbons, are discussed separately.

Organic Industrial Waste

Organic industrial waste of natural origin includes the pollutants listed in Table 4.1. These wastes can be harmful to receiving waters as a result of their toxicity and biochemical oxygen demand and the concentration of suspended solids. Their effects on the marine environment are largely similar to those imposed by municipal wastes, that is, a reduction of oxygen supplies and possibly eutrophication, most likely to occur in semienclosed areas such as bays and estuaries, into which many industrial sources discharge.

Table 4.1. **Some Organic Industrial Wastes of Natural Origin**

Compound	Source
Tannins	Dye industry
Lignin	Pulp and paper mills
Carbohydrates	Pulp and paper mills, brewers, distilleries, sugar production
Proteins, peptides, amino acids, fatty acids, lipids	Slaughterhouses, dairies, fish-processing plants
Pyrenthrines	Insecticides
Terpenes	Flotation of ore

Source. Adapted from Goldberg, 1972.

The toxicity of pulp mill wastes is usually associated with sulfur-containing compounds, arising from the wood digestion process, and with chlorinated phenolic compounds, produced during bleaching of pulp and paper. Untreated pulp effluents have a very high biochemical oxygen demand, and many fish kills have been reported in waters receiving these wastes. In addition, the solid fractions of pulp and paper mill wastes may accumulate in sludge beds and affect benthic communities.

The wastes from breweries and food processing industries similarly exert a high biochemical oxygen demand, which is usually of more practical importance than their direct toxicities.

In the past few decades, the expanding chemical industry has produced countless new synthetic organic chemicals, some of which could pose pollution problems. Thousands of new compounds are developed annually, the result of consumer acceptance and demand for more products such as plastics, synthetic fibers, and polymers. Currently, close to 90 percent of these chemicals are produced directly or indirectly from petroleum.

Given the enormous amounts of synthetic organic chemicals, it is impossible to list all potentially dangerous substances and their impact on the marine environment. Table 4.2 presents some of the more important products and their effects on marine organisms. The list is based on the annual production of these compounds, the percentage of the production reaching the sea, their patterns of use and dispersal, and their toxicity and persistence.

Inorganic Industrial Waste

ACIDS AND ALKALIS

The production of many inorganic chemicals gives rise to large quantities of waste acids and alkalis. Such wastes frequently are discharged into estuaries, and large quantities of acid waste are dumped offshore. In addition, large quantities of acids resulting from the burning of fossil fuels, which releases sulfur dioxide and nitrogen oxides, reach the sea via the atmosphere.

Although these wastes very often have deleteriously affected the pH of many freshwater lakes, the likelihood of this occurring in the sea is small. It is possible that pH changes could be detected over relatively small areas in the vicinity of the discharges but, in view of the sea's large buffering capacity, these changes are shortlived and appear to have little impact. The acute toxicity of both acids and alkalis to some marine animals is documented but, again as a result of the sea's buffering capacity, moderately large concentrations appear to have little impact. When excessive amounts of acid waste are discharged in coastal waters, the effects can be noticeable, however. In polluted areas such as the German Bight in the North Sea, fishermen often bring up fish with tumors or various diseases. It is believed that acid waste disposal contributes to these effects, though the industry of course denies this. As a result of insufficient data, the issue remains unresolved.

CYANIDES

Wastes containing cyanides are discharged to rivers, estuaries, and the sea from a variety of industrial sources, including metal plating plants, coke ovens, and chemical factories. Discharges of HCN, NaCN, KCN, and similar simple cyanides in the sea, lead to the formation of HCN, about one tenth of which is dissociated to CN^-. HCN is very toxic to marine life, but detrimental effects associated with the discharge of cyanides generally are believed to be predominantly local in nature.

HEAVY METALS

Wherever industrial wastes are discharged, heavy metals such as copper, zinc, cadmium, mercury, and lead are found. Some heavy metals are among the most dangerous pollutants as a result of their persistence and toxicity. They reach the ocean through every possible means of industrial disposal: outfalls, rivers, runoff, dumping, and the atmosphere. It should also be noted that substantial amounts enter the sea through natural processes such as erosion and the weathering of rocks.

The effects of heavy metal pollution on marine organisms depend on a wide variety of factors. Consequently, it is difficult to broadly discuss their effects, though some generalizations can be made. All heavy metals, for example, are toxic. It is also known that the toxicity of the metal depends on its physicochemical state. Ionic copper, for example, is much more toxic than copper chelated by organic substances such as humic acids and detergents. Organic mercury and lead, in contrast, are much more toxic than the corresponding inorganic compounds. Toxicity also varies with the species and its stage of development. Generally, less

Table 4.2. Some Organic Industrial Substances and Their Effects on the Marine Environment

Compound	Formula	Use	Pathways	Toxicity (LC 50)	Characteristics and Effects
Acetone	Ch_3COCH_3	Solvent, delustrant, decreasing agent	Rivers, direct discharges	100–1000 ppm	Readily degradable, nonpersistent
Acrolein	$CH_2=CH \cdot CHO$	In production of acrylic and other plastics		<1 mg/l	Unstable and polymerizes rapidly to give disacryl; sublethal effects (decreased shell growth in oysters) reported at less than 0.1 mg/l; repellent to fish
Acrylonitrile	$H_2C=CHCN$	Intermediate in synthesis of acrylic fibers		25–50 ppm	One of the more oxidation-resisting nitriles; relatively short residence time; high cumulative or chronic toxicity to marine organisms
Benzene	C_6H_6	Serves as basic substrate of numerous chemical intermediates in production of petrochemicals and plastics	Ocean dumping, operational discharges, atmosphere, seepage	10–100 ppm	Degrades very slowly; will bioaccumulate and cause tainting
Carbon disulfide	CS_2	Production of rayon, cellulose films; manufacture of tetrachloride and xanthates		<1 ppm	Decomposes in the sea by oxidation; very toxic
Cresol	$CH_3C_6H_4OH$	Manufacture of disinfectants and synthetic resins, tricresyl phosphate, herbicides		1–10 ppm	Biodegradable, breakdown products are less toxic, will taint fish and shellfish at very low concentrations

Crotonaldehyde	CH₃CH:CHCHO	Production of acetate solvents	Industrial effluents; leakage	1–10 ppm	Biodegradable and not accumulated; breakdown products are less toxic
Cumene	C₆H₅CH(CH₃)₂	Intermediate in production of phenol and acetone; manufacture of plastics; solvent	Evaporation	10–100 ppm	Degradable
Ethyl alcohol	C₂H₅OH	Solvent; production of alcoholic beverages; many minor uses		10–100 ppm	Rapidly biodegraded and not accumulated
Ethyl benzene	C₆H₅C₂H₅	Solvent; production of styrene monomer	Industrial effluents, possibly atmosphere	10–100 ppm	Degradable; breakdown products are less toxic; not accumulated
Naphthenic acid	Mixture of carboxylic acids	Production of metal naphthenates		1–10 ppm	Likely to be biodegraded; breakdown products possible with same toxicity; results in tainting of fish and shellfish
Phenol	C₆H₅OH	Chemical intermediate in production of resins, pharmaceuticals, disinfectants, herbicides	Present in most domestic and industrial effluents	10–100 ppm	Although chemically stable, phenol is biodegradable, but the speed of this process in the sea is not known; will react with proteins, but action on marine life has not been studied; tainting may occur at very low levels
Styrene monomer	C₆H₅CH:CH₂	Production of polystyrene and other plastics, rubbers, resins	Chemical tanker washings	10–100 ppm	Degrades very slowly; degradation products of the polymer and monomer in the marine environment are not fully known, but are unlikely to be more toxic than the parent compound

(*Table continued on p. 76.*)

Table 4.2. *(continued)*

Compound	Formula	Use	Pathways	Toxicity (LC 50)	Characteristics and Effects
Phthalate esters		Plasticizers in production of PVC, polyvinyl acetate, and other plastics; solvent, insect repellents, cosmetics		Varies with alcohol used; usually low	Rate of breakdown varies between esters; several phthalates are accumulated by fish and invertebrates by factors ranging from 1400 to 3600
Tetramethyl lead Tetraethyl lead	$Pb(CH_3)_4$ $Pb(C_2H_5)_4$	Antiknock additives	Atmosphere	1ppm (TEL) n.a. (TML)	Short-term bioaccumulation
Toluene	$C_6H_5CH_3$	Solvent; chemical intermediate	Industrial effluents	10–100 ppm	Biological properties are not well known
Toluene diisocyanate	$CH_3C_6H_3(NCO)_2$	Production of polyurethane foams		1–10 ppm	Will react with seawater to produce solid phenyl ureas
Xylene	$C_6H_4(CH_3)_2$	Solvent in manufacture of paints and surface coatings; synthesis of organic chemicals	Spills, tank washings, some via atmosphere	n.a.	Relatively stable, slow breakdown process; will not bioaccumulate

Source. Adapted from National Academy of Science, 1976 and Joint Group of Experts on the Scientific Aspects of Marine Pollution (henceforth GESAMP), 1976.

developed and younger life stages are more susceptible to heavy metal contamination. More highly developed organisms possess better mechanisms for the exclusion of heavy metals or for resisting their toxic effects.

In addition, all heavy metals are persistent. They are, in fact, not destructible and can be accumulated by some organisms to lethal levels (Table 4.3). In some instances, bioaccumulation will not affect the organism itself but rather its predators. Mercury levels in swordfish and some species of whales, for example, are very high, though the high metal burden in these animals is not necessarily the result of pollution.

Marine animals contaminated with heavy metals can present a significant danger to human health. Most people do not consume enough seafood to be seriously affected, but heavy metal contamination can be a problem for those population groups that regularly eat fish or shellfish. This was most dramatically demonstrated in the Japanese fishing town of Minamata. One of the main industries in this small town produced chemicals such as acetaldehyde and vinyl chloride, in which mercury was used as a catalyst. The company producing these compounds discharged its effluents into the sea where the mercury compounds were converted into methyl mercury by microorganisms and subsequently were bioaccumulated by fish and shellfish. Since this process did not discolor or impart an unpleasant taste to the fish, the inhabitants of

Table 4.3   **Bioaccumulation Factors of Heavy Metals in Shellfish**

Metal	Normal Concentration in Seawater (ppb)	Scallops	Oysters	Mussels
Ag	0.1	2,300	18,700	330
As	2.0	2,260,000	83,000	25,500
Cd	0.02	3,000	318,000	100,000
Cu	1	200,000	13,700	3,000
Cr	0.04	80,000	60,000	320,000
Hg	0.005	55,500	80,000	80,000
Mn	2	12,000	4,000	13,500
Ni	5	5,300	4,000	14,000
Pb	0.02	28,000	31,300	4,000
Zn	10		110,300	9,100

Source;   Adapted from Brooks and Rumbsby, 1965, Riley and Chester, 1971, and Papadopoulou and Kaias, 1976.

Table 4.4. Heavy Metals in the Marine Environment

Metal	Use	Pathways	Toxicity (ppm)
Cadmium	Pigments (paints, glass), bearing metals, fusible and other alloys, batteries, nuclear reactors	Effluents, rivers, atmosphere	0.1–100
Arsenic	Arsenious oxide is used in the manufacture of insecticides, wood preservatives, weedkillers, antifouling paints	Metal processing plants, atmosphere	1–10 (arsenite form)
Chromium	Stainless steels, armorplates, certain cutting steels; in chemical industry to produce chromates and bichromates	Rivers, atmosphere	
Copper	Electric wiring, switches, plumbing, plating, roofing, construction, alloys, agricultural chemicals, wood preservatives, pigments	Industrial wastes, mines, atmosphere	1–10
Lead	Storage batteries, antiknock additives, radiation shielding, fusible alloys, pigments, glass, foil	Atmosphere	1–10
Mercury	Chlor-alkali industry; other uses include agricultural chemicals, pharmaceutical products, dentistry, dyes	Atmosphere	0.005–10
Nickel	Alloying element in stainless steels	Rivers, industrial effluents, atmosphere	0.5–10
Zinc	Galvanizing or coating of iron and steel sheets; also in synthetic fibers, batteries, alloys, paints	Rivers, effluents, some atmosphere	1–10

Effects on the Marine Environment

Toxicity to marine animals varies with compounds, species, and stage of development. Marine larvae and algae seem to be most susceptible. Accumulation appears to be via the food chain rather than via direct uptake.

Elemental arsenic is virtually nontoxic to marine organisms. Most of the toxicity data refer to the arsenite form. The acute toxicity level of 1–10 ppm is reported for a wide variety of marine organisms.

The hexavalent form of chromium is considered to be the most toxic and has been suggested as the cause of ulceration in fish. Threshold concentrations vary between 1 ppm for *Nereis* to 40–60 ppm for *Carcinus* and 33 ppm for *Crangon*. Oyster larvae are killed at concentrations of 5–10 ppm and 1 ppm has been observed to reduce the rate of photosynthesis in algae.

Acute toxicity to marine animals is usually reported in the order of a few ppm for adults of numerous species. Larvae are consistently more sensitive—0.01 ppm for sea urchin larvae, for instance. Similar levels have been shown to affect photosynthesis and growth in algae. Fish and crustaceans appear to be able to regulate their body burden; this is not the case for less developed species.

Lead is an enzyme inhibitor and impairs cell metabolism. In marine animals it is assumed that acute exposure damages gill surfaces, thus inhibiting oxygen–carbon dioxide transfer. Mode of uptake is generally through food rather than through direct uptake. Larvae are generally more susceptible.

Acute toxicity of ionic mercury varies with the species and its stage of development. Organic compounds are much more toxic. The figures for acute toxicity are time-dependent, apparently with a very low threshold. Behavioral abnormalities have been observed at sublethal levels.

Toxicity in seawater appears to be lower than in freshwater. Less developed life forms and larval stages are more vulnerable.

Acute exposure leads to gill damage but uptake by marine organisms is primarily via food rather than form seawater. Larvae are much more susceptible than adults, as has been demonstrated with oysters, sea urchins, and other invertebrates. *(Table continues on p. 80.)*

Table 4.4.   (*continued*)

	Effects on Human Health
Cadmium	Cadmium vapor and cadmium salts can induce acute toxic injury to the lungs, gastrointestinal tract, and kidneys, but such injury is unlikely to occur on contact with polluted seawater or through the consumption of contaminated seafood. Cadmium is only slowly excreted by the body and gradually accumulates with age.
Arsenic	Certain inorganic arsenic compounds are known to be carcinogenic and highly toxic on acute or long-term administration. Inorganic arsenic compounds can penetrate the skin but it is unlikely that contaminated seawater could contain a concentration sufficient to cause injury. Organic compounds are more rapidly excreted and are less toxic.
Chromium	High concentrations of chromium salts irritate the lungs, mucous membranes, gastrointestinal tract, and skin. These effects occur, however, only on prolonged contact and could not be caused by polluted water. Amounts found in seafood are harmless to humans since chromium does not accumulate significantly in the body and is poorly absorbed from the gastrointestinal tract.
Copper	Large doses of copper salts produce injury to the gastrointestinal tract and the liver and this may cause death. Exposure to such doses by contaminated seawater or seafood is not possible. Copper does not accumulate in tissues and chronic exposure from marine sources is therefore unlikely to cause injuries.
Lead	The acute toxicity of inorganic lead is relatively low in humans, but when exposure exceeds the excretion rate accumulation will occur, provoking chronic plumbism. There have been no instances of human poisoning from eating contaminated seafood, but the rapid increase of lead concentrations in surface waters and sediments could result in health hazards where shellfish form a substantial portion of the diet.
Mercury	The toxicities of elemental and inorganic or organic mercury compounds are different. Poisoning by elemental mercury affects the nervous system. Large

Mercury (*cont'd.*)	doses of inorganic mercury compounds are concentrated in and damage the kidneys. Long-term exposure to organomercury compounds damages the nervous sytem: this has occurred following the consumption of contaminated seafood (Minamata). The accumulation of organomercury compounds in some marine animals has led to the regulation of fishing and the control of the sale of these animals in some countries.
Nickel	Large doses of nickel salts cause acute gastrointestinal irritation but poisoning in man is almost unknown. Nickel is poorly absorbed and there is no evidence of accumulation in the body. Frequent contact will cause dermatitis, but it is unlikely that ordinary exposure to the low concentrations of nickel in the marine environment consitute a health hazard. Similarly, the consumption of seafood is not likely to be harmful.
Zinc	The ingestion of high levels of soluble zinc salts can cause gastrointestinal upsets but the concentrations needed to produce this effect are much higher than could occur in polluted water. Constant exposure to low concentrations does not cause accumulation in human tissues.

Source.   After GESAMP, 1976.

Minamata, dependent on seafood for a large percentage of their diet, continued to consume large quantities of locally caught fish and shellfish. In the late 1950s, however, the population began to show symptoms of a disease that became known as the Minamata disease. Later it was discovered that the Minamata disease was caused by methyl mercury poisoning. More than 120 persons died and hundreds suffered permanent nervous system damage in this dramatic incident, the first and most famous case of mass mercury poisoning from industrial effluents.

Some of the effects of the most prevalent heavy metals are summarized in Table 4.4. There is a growing body of literature on heavy metal pollution, but the picture is far from complete. Most of the data available indeed are concerned with acute toxicities, leaving many unknowns in the understanding of sublethal, chronic, and synergetic effects. Moreover, most results are based on experiments conducted in the laboratory rather than in the ocean. While this procedure is far more practical, the resultant data may fail to convey exactly what happens in the infinitely more complex oceanic environment. Some data, in other words, may over-estimate the effects; others may very well underestimate the potential impact.

Chlorinated Hydrocarbons

Chlorinated hydrocarbons constitute a class of synthetic chemicals that have come into widespread use in both agriculture and industry since the beginning of this century. They are used for a wide variety of purposes, for example, in aerosol propellants, fumigants, fire extinguishers, hydraulic fluids, refrigerants, solvents, heat transfer agents, and dielectric insulants, and as intermediates in organic synthesis.

Many chlorinated hydrocarbons are quite stable. Chlorination generally increases the stability of hydrocarbons to both chemical and biological degradation, and the resulting long residence times of some chlorinated hydrocarbons can have a significant impact on the marine environment. Fortunately, not all chlorinated hydrocarbons are persistent. In fact, the number of compounds to be taken into account in environmental impact reviews is substantially reduced because the nonpersistent segment is unlikely to be encountered in significant quantities in the sea. The major contaminants include a number of low-molecular-weight chlorinated hydrocarbons, the organochlorine insecticides such as DDT, aldrine, and dieldrin, and the polychlorinated biphenyls (PCBs).

Low Molecular Weight Chlorinated Hydrocarbons

The low molecular weight chlorinated hydrocarbons are mainly used as solvents, decreasing agents, or chemical intermediates. They reach the sea primarily via the atmosphere and, to a lesser extent, by rivers (Table 4.5).

The effects of these compounds on aquatic ecosystems are poorly understood. Most of the literature consists of studies of acute exposures to selected compounds, but there is little information on sublethal effects, synergisms, antagonisms, or additive effects on marine organisms. Generally, many low molecular weight chlorinated hydrocarbons break down more rapidly than the heavier compounds such as the insecticides and the PCBs, and bioaccumulation appears to be low. Toxic thresholds for

Table 4.5. **Toxicity and Effects of Low Molecular Weight Chlorinated Hydrocarbons in the Marine Environment**

Compound	Formula	Use	Pathways	Toxicity (ppm)	Characteristics and Effects
Carbon tetrachloride	CCl_4	Manufacture of aerosol propellants and refrigerants; solvent	Evaporation	10–100	Bioaccumulates in fish and in muds; breakdown products not fully known
Chlorobenzene	C_6H_5Cl	Intermediate in production of phenol from benzene; starting material in dyestuffs manufacture; production of fine chemicals	Atmosphere, may be present at ppm level in effluents	2.5 (96 hr LC 50)	Accumulates rapidly in fish, particularly in the liver. Broken down by microorganisms to hydrochloric acid and phenol, which break down to carbon dioxide and water. May accumulate in sediments
Chloroform	$CHCl_3$	Pharmaceutical products, flavoring, solvent; in manufacture of aerosol propellants and refrigerants	Atmosphere	30 (96 hr LC 50)	Biodegradable; breakdown products are less toxic. Will taint fish and shellfish at very low concentrations
O-dichlorobenzene	$C_6H_4Cl_2$	Manufacture of isocyanates and dyestuffs; solvent, pesticide, decreasing agent	Effluents, evaporation	1	Little information available on biodegradability or breakdown products. Highly toxic to fish. Accumulates rapidly in the liver
P-dichlorobenzene	$C_6H_4Cl_2$	Deodorant, moth repellant, soil fumigants	Evaporation	1.5	Biodegradability and breakdown products not investigated. No information regarding bioaccumulation
Epichlorohydrin	$CH_2\text{-}CHCH_2Cl$	Production of epoxy and phenoxy resins		1–10	Readily broken down by bacteria and other aquatic organisms. Final and intermediate breakdown products are less toxic. Unlikely to be bioaccumulated

(*Table continued on p. 84.*)

Table 4.5. (continued)

Compound	Formula	Use	Pathways	Toxicity (ppm)	Characteristics and Effects
Ethylene dichloride	$CH_2Cl \cdot CH_2Cl$	Solvent for gums, fats, and waxes; used in production of paints, artificial fibers, and plastics	Industrial effluents	+100	Degradable, breakdown products are less toxic. Short-term bioaccumulation
Hexachlorobenzene	C_6Cl_6	Byproduct in manufacture of many chlorinated hydrocarbons	Atmosphere	500–1000	Very stable and unreactive compound. Accumulated but not metabolized to any significance in fish. Single dose for acute toxicity is fairly large but chronic toxicity appears to be significant
Trichlorobenzene	$C_6H_3Cl_3$	Intermediates in dyestuff manufacture	Evaporation	Low in range (<10)	No information on bioaccumulation or breakdown products
Vinyl chloride	$CH_2{:}CHCl$	Used in manufacture of polyvinyl chloride (PVC)	Aerial transport		Does not remain in the sea very long because of low vapor pressure and low solubility. It has been suggested that the monomer may be more harmful than was previously thought, however

Source. Adapted from GESAMP, 1976.

mammals and aquatic organisms are on the order of parts per million, which is about six orders of magnitude larger than known levels in the ocean. As a result, the uses and the release of these substances does not yet appear to present a hazard.

Insecticides

Several chlorinated hydrocarbon compounds such as DDT, Dieldrin, Endrin, and Aldrin (Fig. 4.1) are used on a wide scale as insecticides. These compounds reach the sea primarily through aerial transport, but large amounts are also present in domestic or industrial effluents, or

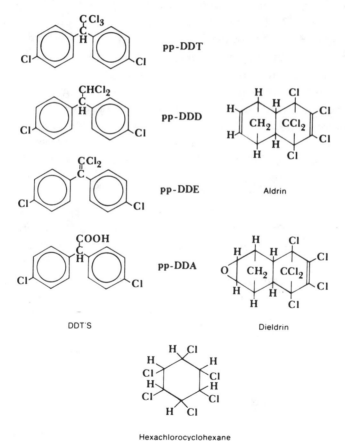

Figure 4.1. Chlorinated hydrocarbons used as insecticides.

will be added to rivers and oceans through runoff. Estimates of the annual proportion of insecticides that reaches the ocean range from 40 to 60 percent.

Once in the sea, organochlorine insecticides are concentrated in sediments and in marine organisms. As a result of their lipophylic/hydrophobic character, they are particularly concentrated in any oily material. They are therefore found predominantly in animals with a high lipid content and animals higher up in the food chain. Residence times for these compounds have not been accurately determined, but they generally range in the order of several years or more.

The exact mode of action of organochlorine insecticides is not fully understood, but it is generally accepted that they affect the transmission of impulses in the central nervous system. Their acute toxicity to marine organisms is well documented and appears very high, particularly in crustaceans. Sublethal side-effects have also been observed, but more frequently in air-breathing organisms such as seals and water birds than among water-breathing organisms such as fish. This appears to be mainly the result of differences in the body chemistries of these groups of animals.

Restrictions on the use of DDT in various countries have resulted in a corresponding increase in the use of other insecticides such as Aldrin and its breakdown product Dieldrin, or toxaphene. These compounds are even more toxic than DDT, but their residence time in the marine environment is considerably shorter.

PCBs

Polychlorinated biphenyls form a group of extremely stable, fire-resistant, chlorinated hydrocarbons. These properties make them ideal insulating fluids in electrical equipment, particularly in high-power transformers and capacitors. They have also been used in paints, sealants, lubricant additives, hydraulic fluids, and heat exchange fluids. Depending on the degree of chlorination of the biphenyl molecule, there are about 200 different PCB formulations (Fig. 4.2).

Losses during the manufacturing process and leakage from products containing PCBs are the two major routes of PCB entry into the environment. Once introduced, PCBs are transported to the ocean primarily through adsorption to fine particles in rivers and, to a lesser extent, by airborne particles. In occurrence and distribution PCBs behave in much the same way as the insecticides and they are, at least in the sea, just as widely spread.

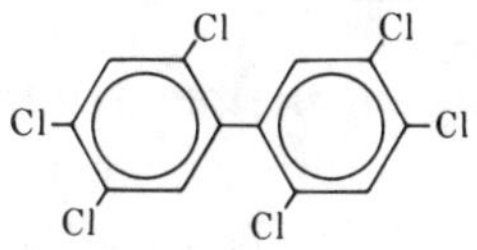

Figure 4.2. Polychlorinated biphenyl (PCB).

PCBs are extremely resistant. Biological decomposition is indeed very slow since they lack the ethane group between the aromatic rings, which is the site of action of most DDT transformations.

They are also very toxic, though their acute toxicity appears to be less than that of the organochlorine insecticides. There is evidence, however, that they have a high chronic toxicity, which simply means that it may take from a few weeks to a few generations before low-level exposure effects appear. There is a considerable volume of literature on the impact of PCBs on the marine environment. Some commonly observed effects include a reduction in the reproductive capacity and in the survival of younger life stages, diminished immunological capacities, and learning and behavioral deficiencies. The mode of action of PCBs is not well understood, though it is assumed to be similar to that of the organochlorine insecticides.

Human health effects due to low PCB levels comprise abnormal fatigue, abdominal pain, numbness, coughing, acne, and headaches. There is some evidence that PCBs are carcinogenic, but there are as yet no known ill effects on humans which can be associated with the levels of PCBs in the marine environment.

In view of the widespread PCB contamination of the environment, the use and production of these compounds have been restricted or banned in several countries. The resulting reduction in inputs has not yet been accompanied by a major decrease of PCB levels in the sea, mainly as a result of their persistence and the continued leakage from PCB-containing products.

Oil

Oil is one of the major sources of pollution in the ocean. It is undoubtedly the most publicized form of marine pollution, mainly as a result of its visibility and its effects on birds—animals that apppear to elicit emotional outcries more readily that do fish or other marine organisms, with the exception of whales, of course. There is little doubt that the outcry is justified since most of the oil that enters the marine environment is introduced by people and could, as a consequence, be reduced. But there is a need for some perspective because tanker accidents and blowouts, which can cause massive oil spills, are neither the most significant nor the most dangerous source of oil pollution, as seems commonly believed.

Accidents, in fact, constitute a relatively minor source of oil pollution. At least that was the conclusion of a study undertaken by the National Academy of Sciences, which estimated a few years ago that every year approximately 6.1 million metric tons of oil enters the ocean. Environmentalists considered this to be a conservative estimate; the oil industry, not surprisingly, contended that it was too large. Regardless of these opinions, the report established an authoritative baseline amount which is probably still fairly representative (Table 4.6).

According to the report, tanker accidents and offshore drilling contributed 200,000 and 80,000 tons, respectively, to the total amount. These figures are now larger: in 1980 tanker accidents led to the release of 390,000 metric tons while two years earlier, the *Amoco Cadiz* single-handedly added 220,000 metric tons of crude oil. Similarly, the contribution from drilling operations has risen: while the 1977 *Ekofisk* blowout released 22,500 metric tons into the North Sea, this seemed small in comparison to the *Ixtoc I* blowout off the Mexican coast, which for a period of 295 days (June 3, 1979–March 23, 1980) spilled over 450,000 metric tons. Even greater amounts of oil are released by less spectacular sources: the operations of ships and runoff from land. Other major sources include refineries, domestic and industrial effluents, and the atmosphere. Natural sources such as submarine seeps contribute more than 600,000 metric tons, while the production of hydrocarbons by marine plants may amount to as much as 3 million tons.

The introduction of such amounts of hydrocarbons is bound to have a certain impact. But any assessment is difficult because oil is composed of thousands of organic compounds, including compounds with straight and branched chains, olefins, saturated and aromatic ring compounds, and a number of heterocompounds that contain oxygen, nitrogen, or sulfur. As could be expected then, the effects of oil and oil products depend on their composition.

When oil enters the sea, it tends to spread relatively rapidly. Some of the lighter fractions will evaporate immediately, while heavier compounds will be dissolved or sink. Most of the dissolved oil will slowly

Table 4.6.   **Source and Volume of Petroleum Hydrocarbons Entering the Ocean**

Source	Quantity (millions of metric tons)	Percentage of Total
Marine transportation		
LOT tankers	0.31	5.07
Non-LOT tankers	0.77	12.59
Drydocking	0.25	4.09
Terminal operations	0.003	0.05
Bilges bunkering	0.5	8.18
Tanker accidents	0.2	3.27
Nontanker accidents	0.1	1.64
Total	2.13	35.23
Other marine sources		
Natural seeps	0.6	9.82
Offshore production	0.08	1.31
Total	0.68	11.13
Land-based sources		
Coastal refineries	0.2	3.27
Atmosphere	0.6	9.82
Domestic wastes	0.3	4.91
Industrial wastes	0.3	4.91
Urban runoff	0.3	4.91
Rivers	1.6	26.17
Total	3.3	53.99
Total	6.113	

Source.   National Academy of Sciences, *Petroleum in the Marine Environment*, Washington, D.C., 1975, p. 6.

be degraded or metabolized by bacteria, but this breakdown depends on a variety of factors, such as the climatic conditions and the original composition of the oil. The oil that sinks will persist for a longer time because deeper waters have lower temperatures and lower oxygen concentrations. In fact, once into the sediments, degradation will nearly come to a halt when the sediments are anaerobic.

The most noticeable effects of oil pollution are on intertidal benthic communities and bird populations. But as in most cases of marine pollution, the problem is not so much one of determining the immediate effects as determining the sublethal and chronic effects. Massive oil spills wipe out a benthic community, or may taint shellfish, but all in all most scientists agree that such a single heavy contamination has a neglible impact in comparison to the worldwide, continuous exposure to smaller concentrations.

When highly dispersed, oil—and particularly the aromatic fractions— is toxic to marine life. The effects may range from the poisoning of certain organisms to the total disruption of the ecosystem, caused by the destruction of the more sensitive younger life stages or the elimination of the food sources of higher species. It is also believed that traces of oil may affect fish behavior by interfering with communication systems. Some oils, in addition, contain carcinogens and their occurrence in marine animals could present a health risk to humans. And last but not least, oil pollution, more than any other source of marine pollution, interferes with other ocean activities such as fishing, mariculture, and particularly recreation.

When substantial amounts of oil are spilled, several recovery methods can be used to reduce or restrict the damage. Mechanical methods are available to remove oil from the surface of calm waters, but no really adequate technique exists for handling oil spills in rough waters. Detergents can be used to disperse the oil. This technique was applied to control the spread of oil following the 1967 *Torrey Canyon* grounding, though the detergents, highly toxic themselves, caused far greater mortality among marine animals than the oil itself. Oil dispersants currently available are of low toxicity and are more effective at low concentrations. Combustion of spilled oil is not a bad method for treating an oil spill, but it is usually very difficult to burn the oil, unless this is done immediately after the spill. Finally, the development of oil-degrading bacteria strains represents another alternative. Some strains have been produced that digest the oil relatively rapidly, but it appears that some of the intermediate products are dangerous. It should be observed, however, that all of these removal techniques can only reduce the impact of accidental spills, which are a relatively minor component of the total amount of oil that annually enters the sea. Reduction of the major sources—vessels, effluents, and runoff— requires preventive measures, but often these are even more difficult to apply.

Radioactive and Thermal Waste

Radioactive substances are naturally present in the marine environment (Table 4.7). In addition, substantial amounts of radioactive materials are

Table 4.7. **Principal Naturally Occurring Radionuclides in the Marine Environment**

Nuclide	Half Life (years)	Concentration (g/l)
^{3}H	1.2×10	3.2×10^{-18}
^{10}Be	2.7×10^6	1.0×10^{-13}
^{14}C	5.5×10^3	3.1×10^{-14}
^{32}Si	7.1×10^2	
^{40}K	1.3×10^9	4.5×10^{-5}
^{87}Rb	5.0×10^{10}	3.4×10^{-5}
^{226}Ra	1.6×10^3	8.0×10^{-14}
^{228}Th (RdTh)	1.9	4.0×10^{-18}
^{228}Ra (MsTh)	6.7	1.4×10^{-17}
^{230}Th (Io)	8.0×10^4	6.0×10^{-13}
^{231}Pa	3.2×10^4	5.0×10^{-14}
^{232}Th	1.4×10^{10}	2.0×10^{-8}
^{235}U	7.1×10^8	1.4×10^{-8}
^{238}U	4.5×10^9	2.0×10^{-6}

Source. Adapted from Rice and Wolfe, 1971.

introduced artificially, particularly through the fallout from nuclear explosions and through the use of nuclear energy facilities, including reprocessing plants.

The input of radioactive material into the ocean from past nuclear explosions does not appear to have produced major problems. Nuclear tests from 1952 to 1958 produced more than 4.5 million tons of fission products, most of which fell out over the oceans, but this amount was distributed fairly evenly and consequential concentrations in the sea remained very small, with the exception of these areas very near the testing sites. Uncontrolled inputs of artificial radioactivity still occur as a result of these tests, but the rate of input is now much smaller than the rate of loss due to decay.

The amount of radioactivity entering the oceans from nuclear power plants, subject to strict controls, is lower than the contribution of atmospheric fallout. This source of pollution tends to be more localized, however, and therefore more dangerous.

Radionuclides can be accumulated by marine organisms (Table 4.8) by highly selective ion transport mechanisms, physical adsorption pro-

Table 4.8.   **Approximate Enrichment Factors for Radioisotopes of Probable Significance in the Marine Environment**

Radionuclide(s)	Algae	Crustacea	Molluscs	Fish
^{3}H	0.90	0.97	0.97	0.97
^{7}Be	250	—	—	—
^{14}C	4,000	3,600	4,700	5,400
^{24}Na	1	0.2	0.3	0.13
^{32}P	10,000	20,000	6,000	37,000
^{45}Ca	2	120	0.4	1.2
^{46}Sc	1,200	300	—	750
^{51}Cr	2,000	100	400	100
^{54,56}Mn	3,000	2,000	10,000	200
^{55,59}Fe	20,000	2,500	10,000	1,500
^{57,58,60}Co	500	500	500	80
^{65}Zn	1,000	2,000	15,000	1,000
^{85}Kr	1	1	1	1
^{89,90}Sr	50	2	1	0.2
^{90,91}Y	500	100	15	10
^{95}Zr^{95}Nb	1,500	100	5	1
^{103}Ru, ^{106}Ru, ^{106}Rh	400	100	5	1
^{110}Ag	—	7	10,000	—
^{132}Te, ^{132}I	—	—	—	—
^{131}I	5,000	30	50	10
^{133}Xe	1	1	1	1
^{137}Cs	15	20	10	10
^{140}Ba, ^{140}La	25	—	—	8

Source.   Adapted from Rice and Wolfe, 1971.

cesses, or the ingestion of previously contaminated organic matter. Some of these substances are toxic but there is generally more concern for the problems associated with exposure to radiation than for problems of toxicity and bioaccumulation. Radiation indeed can lead to somatic and genetic changes. There are great variations among different species as to their resistance to somatic changes but, as is the case with other pollutants, different life stages generally vary in sensitivity. The production of genetic changes in marine organisms remains without definite evidence.

The major danger of radioactive pollution is not of polluting the entire ocean, but of damaging a small restricted area where the material cannot be dispersed rapidly. With strict controls and regulations, this possibility can be minimized but accidents can always happen, and the chance of mishaps increases as the number of nuclear power plants throughout the world increases.

Another problem concerns the disposal of spent nuclear fuel. There are very few alternatives for the disposal of this accumulating high-level waste, and the deep sea bottom has been suggested as one potential disposal site. If this were to be implemented, it could become another highly localized source of radioactive marine pollution. Low-level radioactive wastes have been deposited in the sea for a long time, particularly off the East and West coasts of the United States in the past (and possibly again in the future), and in the Bay of Biscay by a number of Western European nations (particularly the United Kingdom, Belgium, the Netherlands, and Switzerland). This is, of course, an extremely sensitive and controversial issue. While these wastes do not appear to pose an immediate threat to marine life and humanity, their disposal in the sea strikes many as irresponsible. The problem is that there are not many alternative sites (on land) that would be either more acceptable or less dangerous. At any rate, the dumping of low-level radioactive wastes calls for considerable additional research on its potential impact and alternative disposal methods, as well as strict monitoring.

Thermal pollution results from the large volumes of water needed to cool power plants. Most of these plants are located near water bodies such as lakes, estuaries, or the sea, from which they take water for cooling purposes and in which they discharge it afterward. Usually, the temperature of the water is 10 to 20°C higher than it was initially, and this can have a significant local impact.

The two major effects of thermal pollution are a decrease in oxygen solubility and an increase in the metabolic activities of aquatic organisms. This may lead to a higher biological oxygen demand, mortalities of both larval and adult animals, and changes in biological communities. Although thermal pollution has not yet caused any serious problems in the marine environment, the anticipated increase in the number of nuclear power plants, which produce as much as 50 percent more thermal waste per kilowatt-hour of capacity, will substantially increase the demand for cooling water. Special care is therefore needed in siting these plants, particularly in tropical areas where many aquatic plants and animals are extremely sensitive to temperature anomalies.

4.2.   THE STATUS OF MARINE POLLUTION

The seas of the North American continent are affected by pollutants representing the needs of an industrial society. The consequences of this pollution vary according to the pollutants and the geographic location of the source. The Arctic regions, for instance, are more fragile than temperate or tropical regions because low temperatures reduce the receiving capacity of the water. The oil reserves within the Arctic Circle, and the possibility of spills, thus constitute a grave threat to this area.

Major polluters in Canada and Alaska are fish processing plants and pulp mills. The organic wastes produced by these industries reduce the oxygen content of the water and, as a result, the chances of survival of young fish. Insecticides have taken a toll as well: before the restrictions on the use of DDT, for instance, land runoff caused tremendous mortalities (50–98 percent) among young salmon. Domestic waste, 20 percent of which is still discharged untreated, resulted in the closing of well over one fourth of the coastal shellfish grounds in this area.

In the United States, more than 40 percent of all manufacturing plants are located in the coastal zone, with a marked concentration in the Mid-Atlantic region and the Great Lakes, and more recently the Gulf Coast. Industrial pollutants, as a result, are the major source of pollution in U.S. coastal waters. Other sources include domestic waste, a relatively large proportion of which is discharged without secondary treatment, and thermal waste. These sources have contributed to a noticeable decrease in the productivity of these coastal waters.

Marine pollution in European coastal waters involves more or less the same problems. The physical characteristics of the Baltic Sea make it an area that is very vulnerable to pollution. Its sewage load is expected to increase but the treatment of wastes from paper and pulp mills has been improved considerably. Mercury remains one of the contaminants of greatest concern in this region, and fish sales have been banned at times because of the high concentrations found in fish. In addition, Baltic fish, seals, and birds have shown high levels of chlorinated hydrocarbons such as PCBs. Some persistent insecticides and the use of mercurials for seed sterilization have been banned, however, and this has resulted in a gradual decline in the circulation of these pollutants in the Baltic Sea. Oil tankers of up to 100,000 dwt can enter the Baltic, so that oil spills are a permanent threat. Hundreds of smaller oil spills are recorded annually, but severe cases have been very few thus far.

The North Sea, one of the richest fishing regions in the world, receives wastes from several countries with high populations and great industrial activity. Large quantities of sewage are discharged in rivers and coastal outlets as well as through pipelines, which move the pollution load from estuaries to offshore waters. The levels of pesticide residues in some fish species are high, but as the use of such substances in the bordering countries is decreasing, it can be assumed that no further increase will occur. In view of the large quantities of oil transported through the North Sea, accidental as well as operational oil pollution remains a constant threat.

The dominant problems in the Mediterranean are sewage, most of which is discharged without much treatment, and oil. In Italy, for instance, only 32 out of 8049 towns had complete sewage treatment facilities during the early 1970s. The western Mediterranean receives the bulk of industrial pollution, derived mainly from the food processing and chemical industries in Spain, France, Italy, and Algeria.

Severe marine pollution problems also exist in eastern European countries, particularly since high priority has been given to rapid national growth, leaving environmental concerns somewhat neglected. The Caspian Sea has been an obvious victim of this policy: during the last 40 years there has been a significant reduction in its fish catches, the main causes being industrial pollution and the construction of hydroelectric power plants along the Soviet side of the coast.

Highly industrialized Japan, too, suffers from virtually every type of marine pollution, a process which has developed dramatically since the early sixties. Coastal marine pollution is on the increase, and has caused considerable damage to fisheries, particularly in the Seto Inland Sea, which is an important mariculture region. Some of the most important sources of pollution include sewage, pulp mills, the fermentation industry, fish processing plants, synthetic fiber plants, the steel industry, and refineries. The largest coastal areas affected are in the Tokyo, Osaka, and Ise Bay regions, which receive large quantities of mixed effluents.

There is not much quantitative information on the marine pollution situation in Australia, but it appears that the coastal region in the vicinity of 12 population and industrial centers are significantly affected. In addition, Australia has a significant potential for offshore oil production, and exploration continues in a number of areas, including the unique Great Barrier Reef.

The marine pollution situation in the developing countries is not very clear, and data on it remain scarce. The problems confronting developing countries are largely the same as those found in the coastal waters of developed nations: sewage, insecticides, and increasing amounts of industrial pollutants. Oil pollution is a major problem in the Southeast Asian archipelago, along the coasts of Africa, and along the northern coast of Latin America. Usually, none of the bordering countries are prepared to combat this problem effectively. In view of the high priority developing countries give to economic development, increases in coastal marine pollution can be expected.

While this brief overview raises justified concerns over the deterioration of some coastal areas, it appears that the open ocean is in better condition. At least that was the conclusion of a recently published, much publicized report on the *Health of the Ocean*, prepared by the Joint Group of Experts on the Scientific Aspects of Marine Pollution (GESAMP). Over the four years of study and research leading to the report, no serious effects were detected in the open ocean that could be attributed to pollution. This does not mean, of course, that there is no reason for concern about marine pollution in the open ocean. Instead, the members of GESAMP confirmed the deterioration of many coastal areas, a process which—if allowed to continue—might spread to the open ocean.

4.3. THE REGULATION OF MARINE POLLUTION

Marine pollution is controlled by a great variety of laws and regulations, most of which were adopted in the late 1960s and early 1970s after the grounding of the *Torrey Canyon* focused worldwide attention on the possibility of polluted oceans. But before venturing into a brief discussion of these various instruments, it may be of interest to review the basics of the economics of pollution. And this review will sound familiar because many of economic problems of pollution are similar to those encountered in fisheries.

In the section on fishery economics, it was observed that a competitive market economy does not function well in matters concerning the allocation of common property resources—those resources that can be used by more than one individual and over which no single user has exclusive rights. Common property resources, it was indicated, are not allocated properly, either because externalities come into play or because the price system fails to convey correct information about their relative scarcity.

For these very same reasons, the private market system will often produce harmful spillover effects on the environment. A firm that dumps its waste into the ocean, for example, does so to minimize its costs. That fishermen have to increase their effort to maintain the same catch is external to the firm, and therefore not really its concern. Moreover, the firm imposes a cost on society because a clean ocean is something we value and something the firm's waste disposal affects, but again it is not always too concerned about this.

The polluting firm could reduce its impact on the ocean by treating its effluents before discharge or by reducing production, but why would it do so unless required? Any such option would indeed increase its costs and lower its competitiveness. As a result, the effluents, in the absence of controls or regulations, are simply discharged into the ocean. If only one firm were involved in this, there wouldn't be much of a problem, but with every factory, or town, or even country following this philosophy, some very serious problems emerge.

Treating the ocean as a common good will lead not only to overfishing but also to pollution. In the case of fisheries, the situation can be rectified by implementing regulations which, in essence, eliminate the common property nature of fish stocks. Marine pollution can be reduced by similar methods. Common approaches to environmental problems include taxation schemes, such as effluent charges, or regulations that specify a certain limit on waste production. All of these methods attempt to internalize the cost of pollution by letting the polluter pay. In fisheries, the only way to prevent waste and misuse is enclosure; similarly, marine pollution can only be reduced by assigning a value to the ocean and requiring some type of payment for its use as a waste dump.

Marine pollution control, if it is to be effective, is a matter of international politics. Pollution is indeed international in its sources and effects: all nations pollute the sea and the impact is not confined locally. As a result, it makes sense to address marine pollution issues by means of international agreements, even though such instruments have a number of shortcomings which, particularly in an area of rapid flux such as marine waste disposal, reduce their efficacy.

As is the case with respect to the regulation of other marine activities, the 1958 Geneva Law of the Sea Conference established a basis for marine pollution regulations, even though the four resultant conventions remained rather vague. The Convention on the High Seas, for instance, required states to draw up regulations to prevent oil and radioactive pollution, while the Convention on the Continental Shelf pointed out that the exploitation and exploration of the shelf should not interfere with the conservation of fisheries. Both conventions were too general to control marine pollution, so this task was left to more specific agreements, whose development the conventions encouraged.

Throughout the past few years, a considerable number of such agreements have been adopted and implemented, though usually in response to particular problems rather than in anticipation of them. As a result, the international regulation of marine pollution is a diffuse and complex matter. Perhaps the best way to briefly review this complicated body of law is by source: vessels and platforms, dumping, land-based sources, and the atmosphere.

Vessels and Platforms

Ships and, to a lesser extent, platforms are significant sources of oil pollution. This was realized early on and in 1954 a first international agreement on oil pollution was drafted: the International Convention for the Prevention of Pollution of the Sea by Oil. Until late 1983 this agreement, which was amended in 1962 and 1969, provided the basis of the international regime to control oil discharges from ships.

The 1954 Convention and its amendments were essentially concerned with intentional oil discharges—that is, discharges from tanks and engine rooms—still the largest source of oil pollution. The 1954 Convention itself went no further than prohibiting the discharge of these wastes in certain areas, something which was soon found not to be entirely satisfactory. Many tankers, for instance, before arriving in port tended to discharge their oily water ballast and oily tank cleanings into the sea, a practice which led to enormous amounts of discharges since whatever oil remaining in the tanks after unloading (usually about 0.33 percent of the original cargo) was thus flushed out into the ocean.

To restrict these discharges, the Convention was amended in 1962. Subsequent to the new rules, the industry developed the Load on Top (LOT) system, whereby tank washings were collected in a slop tank (Fig. 4.3). During the voyage, the oil in the slop tank separated from the cleaning water, which could then be pumped out while the remaining oil was removed in port or incorporated in the next cargo. This represented quite an improvement since most of the world tanker fleet adopted the system, but in many instances the water in the slop tank still contained

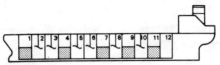

After discharge of cargo from the cargo tanks, sea water is taken on as ballast into some of the tanks (Nos. 1, 4, 7, 9 and 11). Other tanks are washed with sea water (Nos. 2, 3, 5, 6, 8 and 10).

The cargo tank washings are transferred through 'stripping' pipes to a slop tank (No. 12), (which may be just another cargo tank). These washings settle in the slop tank.

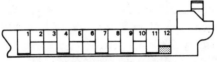

Clean ballast water is pumped into the cleaned tanks (Nos. 2, 3, 5, 6, 8 and 10).
  The dirty ballast water taken on at the start has by now partially settled, so the lower layer of water is discharged to sea (from Nos. 1, 4, 7, 9 and 11), and

the residual oil and water is stripped to the slop tank (No. 12) to settle.

After settling in the slop tank the water layer is discharged to sea (from No. 12) or to port reception facilities when available. A new cargo is loaded on top of the slop tank oil residues (not shown).

Figure 4.3. The Load on Top procedure. (From Royal Commission on Environmental Pollution, 1981.)

excessive amounts of oil. Accordingly, the 1969 amendments were drafted, specifically defining the amounts of oil that could be discharged.

Despite the many good intentions of the Convention and its amendments, not all that much improvement has been observed in the amounts of oil discharged operationally by ships. Perhaps the amount of oil discharged per ton transported has been reduced, but oil traffic has risen markedly over the years, easily offsetting any such improvements. More important, however, is the fact that the provisions are not well enforced. The Oil Pollution Convention and its amendments are administered by the International Maritime Organization (IMO), a United Nations agency concerned with maritime matters, but it has no jurisdiction over enforcement. Only the flag state can do this while other states are restricted to reporting violations to the flag state, unless the violation occurred within their jurisdiction. Many countries, particularly the so-called flag of convenience countries, have not shown much of a commitment to enforcing the rules or penalizing any violation. As a consequence, large amounts of oil are still intentionally discharged into the sea.

In addition, there is also a chance that accidental oil releases occur, and the grounding of the *Torrey Canyon* in 1967 called attention to the fact that international law was not prepared to deal with such matters. Since an oil spill could involve several nations, as the *Torrey Canyon* and its successors made amply clear, it was thought best to regulate accidental oil pollution by treaty. The first such agreement was the International Convention relating to Intervention on the High Seas, concluded in Brussels in 1969. The Public Law Convention, as this treaty is known, permits coastal states to take any measures on the high seas necessary to prevent or eliminate danger to their coastline or related interests, following a serious oil spill. This permits a coastal state, for instance, to bomb and sink a disabled oil tanker to avoid massive pollution, which is exactly what the Royal Navy had done in the case of the *Torrey Canyon*.

Prior to the *Torrey Canyon*, international regulations to prevent accidents, rather than treat them, were virtually nonexistent. A 1971 amendment to the 1954 Convention for the Prevention of Pollution of the Sea by Oil addressed this issue, specifying a number of construction standards for new tankers designed to reduce the size of oil tanks and the rate of escape in the event of an accident, but these provisions never became internationally binding.

Despite these precautions, oil pollution is still liable to cause damage. Since there was a need for some legal machinery to allocate responsibility for compensation, the International Convention on Civil Liability for Oil Pollution Damage was signed in Brussels, also in 1969. This treaty, which became known as the Private Law Convention, seeks to establish international rules and procedures for determining liability and providing compensation for the damage caused by a spill. The money for compensation is supplemented by the 1975 International Convention on the Establishment of an International Fund for Oil Pollution Damage, and by two voluntary compensation schemes established by the tanker industry: TOVALOP and CRISTAL. The provisions and relationship of these various agreements are outlined in Table 4.9.

By the early 1970s, it was realized that the different vessel pollution prevention measures were not satisfactory. Technological developments, such as the increasing size of tankers, required constant updates by technical rules, but it took such a long time to implement new provisions that most were outdated by the time they entered into force. In addition, there was increasing concern, particularly in the United States, about the pollution of the sea from ships by all harmful substances, not only oil. As a result, the International Marine Pollution Conference, which met in London under the auspices of IMO in late 1973, adopted a new Convention on the Prevention of Pollution from Ships, better known as MARPOL 73/78. This agreement, which was amended by Protocol in 1978, superseded the 1954 Convention and its amendments when it entered into force on October 2, 1983, ten years after its adoption.

The far-reaching provisions of MARPOL 73/78 are by no means unrelated to its slow acceptance by the international community. The Convention indeed does not limit itself to oil pollution but also regulates the discharge of noxious liquid substances, harmful packaged goods, sewage, and garbage. In addition, it applies to all types of vessels, including platforms, and covers intentional as well as accidental discharges.

MARPOL contains five annexes, the first of which deals with the prevention of oil pollution. The oil discharge criteria prescribed in the 1969 amendments to the 1954 Convention are largely retained, though the maximum permissible amount of oil that may be discharged is reduced from 1/15,000 to 1/30,000 of the cargo-carrying capacity for new oil tankers. These criteria apply equally to persistent and nonpersistent oils. All oil-carrying ships are required to be capable of operating the LOT system while new tankers need to be fitted with segregated ballast tanks.

The control of pollution by noxious liquid substances is provided for by Annex II. The regulations classify harmful substances into four categories, depending on the degree of danger they represent to the marine environment. More than 400 substances have been evaluated and are included in the annex. The most dangerous substances may be discharged only through reception facilities, whereas other substances can be discharged into the sea, but only in accordance with detailed conditions. Because of the many problems encountered in getting states to comply with these provisions, implementation of this annex was postponed until 1986.

Table 4.9. **Compensation Agreements Relating to Damage by Oil Pollution**

	International Convention on Civil Liability for Oil Pollution Damage (Private Law Convention)	International Convention on the Establishment of an International Fund for Oil Pollution Damage	TOVALOP (Tanker Owners' Voluntary Agreement Concerning Liability for Oil Pollution Damage)	CRISTAL (Contract Regarding an Interim Supplement to Tanker Liability for Oil Pollution Damage)
Purpose	Establishes a uniform international regime under which owners of ships carrying oil in bulk have strict liability for pollution damage resulting from the discharge or escape of oil	Supplements the Private Law Convention to assure adequate compensation to parties suffering pollution damage Indemnifies tanker owners for part of their liability under the Private Law Convention	Provides that tanker owners will compensate persons (including governments) who sustain pollution damage or take preventive measures to mitigate such damage	Supplements the Private Law Convention, TOVALOP, or other sources of compensation to assure adequate compensation to parties suffering oil pollution damage
Status	International treaty; in force 19 June 1975	International treaty; in force 16 October 1978	Agreement among tanker owners: in operation since 1969 as amended 1 June 1978	An agreement among cargo owners in effect since 1971 and most recently amended 1 June 1978
Scope	Seagoing vessels of any type carrying oil in bulk Applies to pollution damage to a contracting state, regardless of where the spill occurs. The escape must come from a ship flying the flag of a contracting state or using its facilities	Contracting States territory and territorial seas although discharge may have occurred elsewhere Vessels flying the flag of contracting states to the Private Law Convention	Seagoing tankers whose owners or charterers are parties to TOVALOP Agreement Applies to pollution damage except when the Private Law Convention applies to the damage; also applies in case of a threat of a discharge, even if no such discharge occurs	Territory or territorial seas of any state, regardless of location of discharge Tanker must be owned (or chartered) to a party to TOVALOP

Damages	Loss or damage by oil contamination including costs of preventive measures and losses caused by preventive measures	Pollution damages not adequately compensated by Private Law Convention because of: No Private Law Convention liability, Financial incapacity of the vessel owner, Damages exceed Private Law Convention limits	Loss or damage by oil contamination or threat of contamination, including cost of preventive measures	Pollution damage not otherwise recoverable from tanker owner or any other source
Liability	$171/Convention ton—not to exceed $18 million per incident	Maximum $58 million aggregate with Private Law Convention Compensation, if any; can be increased up to $116 million by the Assembly of the Fund Indemnifies owner for Private Law Convention liability	$171/Convention ton; maximum $18 million per incident	Maximum $36 million aggregate with all other sources of compensation, if any; can be increased up to $72 million by the Institute Indemnifies owner for part of Private Law Convention liability
Defenses	War, Act of God, intentional act or omission by third party, negligence by government	War, no proof of ship-source spillage, intentional or negligent act of claimant	War, Act of God, intentional act or omission by third party, negligence by government	War, Act of God, intentional act or omission by third party, negligence of governments or claimant
Administration	Government agencies of contracting states	Fund Convention Secretariat, Executive Committee and Assembly (comprising representatives of all contracting states)	International Tanker Owners Pollution Federation	Oil Companies Institute for Marine Pollution Compensation

Annex III contains regulations for the prevention of pollution by harmful substances in packaged forms or freight containers, such as radioactive materials. The government of each contracting state is obliged to issue detailed requirements on packaging, labeling, documentation, stowage and other aspects to prevent or minimize pollution by such substances.

Annexes IV and V provide regulations for the prevention of pollution by sewage and garbage. Under the provisions of Annex IV, ships are not permitted to discharge sewage within 4 miles of land unless they have an approved treatment plant in operation. Between 4 and 12 miles, sewage needs to be comminuted and disinfected before discharge. Annex V sets specific minimum distances for the disposal of all the principal types of garbage and prohibits the disposal of all plastics.

Again, the main problem with an agreement like MARPOL will be its enforcement. The treaty is not self-executing so that each party must enact enabling legislation. In addition, the convention supports the principle of flag state jurisdiction, which proved to be unsatisfactory in the past. On the other hand, the fact that MARPOL 73/78 regulates all types of vessel source pollution, not just oil, and the inclusion of a much more effective amendment procedure, should make it a much stronger instrument for preventing pollution from ships than any of the previous agreements. Furthermore, it should be noted that MARPOL 73/78, like most IMO conventions, provides for a certain degree of port state control and enforcement. The implementation of these provisions, as currently occurring in the United States as well as Western Europe, can also contribute considerably to the effectiveness of the new vessel source pollution prevention regime.

Until MARPOL comes into force in late 1983, pollution from ships other than oil is regulated by a variety of instruments. Pollution from the transport of radioactive material, for instance, is controlled by the Regulations for the Safe Transport of Radioactive Material, drawn up by the International Atomic Energy Agency, and by the 1974 Convention on the Safety of Life at Sea, which seeks to limit the risk of contamination by making provisions for packaging and labeling radioactive goods. Pollution from nuclear-powered vessels is also covered by the Convention on the Safety of Life at Sea, though the provisions do not apply to government-owned vessels, such as warships, which form the bulk of the current nuclear-powered fleet. Another agreement, the Convention on the Liability of Operators of Nuclear Ships, concluded in Brussels in

1962, seeks to impose strict liability for all nuclear damage caused by nuclear ships, but it is not likely to enter into force in view of its unpopularity with many countries.

Dumping

The dumping of wastes at sea is another major source of marine pollution. Virtually every coastal state routinely dumps into the ocean a variety of waste products, particularly dredge spoil but also chemical and other industrial wastes, and until relatively recently there was absolutely no international framework to control this source of pollution.

In fact, the first measure to control dumping was made on a regional basis. In 1972, the countries bordering the North Sea concluded the Convention for the Prevention of Marine Pollution by Dumping from Ships and Aircraft. This agreement does not seek to prohibit dumping entirely but instead attempts to ensure that dangerous wastes, such as very toxic and persistent substances, will not be dumped while others will be disposed of in an appropriate manner. In order to do this, waste products are categorized in three annexes. Products on the so-called Black List cannot be dumped at all, and all other substances require a permit.

Shortly thereafter, it was decided to extend a similar agreement to all the oceans by the 1972 London Convention on the Prevention of Marine Pollution by Dumping of Wastes and Other Matter. The London Dumping Convention, as this agreement is generally known, defines dumping as any deliberate disposal at sea of wastes or other matter from vessels, aircraft, platforms, or other man-made structures at sea, and any deliberate disposal at sea of vessels, aircraft, and artificial structures. The disposal of wastes derived from normal operations and of matter for purposes other than the mere disposal thereof, are not included in this definition, nor is the disposal of wastes arising from the exploitation of sea bed minerals. As is the case in the first agreement, materials that are dumped are divided into three annexes. The disposal of substances in the first annex is completely prohibited, whereas less dangerous substances in Annexes II and III require permits. The London Dumping Convention is somewhat stricter than the earlier agreement because it includes oil, materials produced for biological and chemical warfare, and high-level radioactive waste in its list of prohibited substances, but its is not always taken to heart by all its signatories. It came into force in August 1975.

Land-Based Sources   *Land of Sea*

Although domestic and industrial effluents and the waste loads of rivers are by far the greatest source of marine pollution, there is no global international regime to control this. The reasons for this unfortunate situation are fairly clear: the main effects of land-based pollution are likely to be felt in the coastal waters of the country where the pollution emanates and, more significantly, any international agreement would infringe upon states' rights to control domestic pollution, which many countries do not want.

Even so, in recent years increasing attention has been paid to regional agreements to control land-based sources of marine polution. Pollution from the land is indeed increasing and its effects are no longer merely confined to national waters, spreading instead to the coastal waters of other states or, beyond that, to the high seas.

A number of regional agreements have therefore been concluded to control the waste load of international drainage basins such as the Rhine, which is the largest single cause of pollution in Dutch coastal waters. More recently, a number of treaties have been concluded to regulate land-based source pollution in enclosed or semienclosed seas such as the Baltic Sea and the North Sea. The Regional Seas Programme of the United Nations Environment Programme (UNEP) also contributes to these efforts, particularly in the Mediterranean and in developing nations.

Atmospheric Pollution

As is the case with respect to land-based sources of marine pollution, there are no international agreements covering the substantial contribution of the atmosphere to pollution of the oceans. Atmospheric pollution is very difficult to regulate, however. Any international treaty would involve some interference with domestic pollution regulation, in this instance industrial and vehicle emissions, and states are not likely to give up this type of jurisdiction.

On the other hand, there may be no need for an international agreement, at least not in areas of concern for the oceans at present. Despite the magnitude of atmospheric marine pollution, the pollutants are usually evenly deposited and sufficiently dispersed to prevent the occurrence of harmful effects, though it is a well-known fact that the sea's lead concentrations have increased several times as a result of vehicle emissions. Harmful effects certainly occur on land, where atmospheric pollution has caused the acidification and eventual destruction of many lakes, particularly in Scandinavia and the northeastern United States. But even in such instances, where the damage is evident, the negotiation of regulations or provisions to reduce this input appears extremely difficult and has thus far resulted in few workable arrangements.

Chapter Five

Transportation on the Sea

Navigation for commercial and military purposes is one of the oldest and foremost uses of the sea. Most of the world's trade indeed moves on ships, and in the naval arena the struggle for supremacy over the seas has set the course of world history and continues relentlessly between East and West. Though navigation for commercial and military purposes are intrinsically linked, only merchant shipping is discussed in this section.

5.1. THE STATUS OF MERCHANT SHIPPING

The structure and size of the world merchant fleet as well as those of some of the leading maritime nations are presented in Table 5.1. From these figures it appears that some of the major trading countries do not have very large fleets, but the numbers are somewhat deceptive. Many companies or individuals indeed maintain their vessels under foreign registry and this explains why such small countries as Liberia and Panama have such impressive merchant fleets. Between 35 and 40 percent of the Liberian fleet, for example, is under American ownership, which explains

Table 5.1. **Merchant Fleets of the World**

Country	Number	Gross Tons (thousands)	Deadweight Tons (thousands)
Liberia	2,189	70,718	139,250
Japan	10,652	41,594	68,528
Greece	3,501	40,035	70,232
Panama	5,032	32,600	54,800
Soviet Union	7,713	23,789	26,627
United Kingdom	2,826	22,505	35,990
Norway	2,409	21,862	38,761
United States	6,133	19,111	28,926
France	1,171	10,771	18,728
Italy	1,163	10,375	17,044
China (including Taiwan)	1,619	10,282	15,195
Spain	2,635	8,131	13,879
West Germany	1,782	7,707	12,355
Singapore	849	7,183	12,195
India	644	6,213	10,059
Brazil	666	5,678	9,410
Korea (South)	1,652	5,529	8,882
Netherlands	1,228	5,393	8,430
Denmark	1,152	5,214	8,143
Saudi Arabia	347	4,302	7,769
All other countires	19,788	65,750	96,777
World total	75,151	424,742	701,980

Source. Adapted from Lloyd's Register of Shipping, *Statistical Tables 1982.*

where the American fleet—by far the largest in the world at the end of World War II—has taken itself. In fact, as of late 1981, close to 55 million deadweight tons (dwt) was registered by American shipowners under a foreign flag, a total which makes an interesting comparison with the size of the current U.S.-flag fleet.

There are many advantages to foreign registry, particularly in the so-called flag of convenience countries. The most important, of course, is of a financial nature. Foreign registry is a lot less expensive than national registry, which, as is the case in the United States, may require shipowners to build their ships in expensive domestic yards, staff them with expensive crews, and pay higher taxes. In addition, there are certain legal advantages to foreign registry. International law recognizes the nationality of a ship by its flag, not by the nationality of its owner or its true home port. Since several flag of convenience countries are less strict in controlling and enforcing international shipping and pollution-prevention regulations, vessels or crews that would elsewhere not be considered seaworthy may be able to operate under flag of convenience registry. This situation, not surprisingly, has led to a disproportionate number of such vessels involved in shipping violations and accidents (Table 5.2). Some flag of convenience countries have taken their responsibility somewhat more seriously in recent years, but this does not mean that the situation is being solved. Instead, countries such as Somalia, Cyprus, Honduras, and the Bahamas, eager to attract shipowners, offer lenient registry and may eventually assume Liberia's disreputable safety record.

Ocean shipping is highly dependent on world trade. Throughout this century, the volume of world trade has been expanding continually, as has the size of the world's merchant fleet (Table 5.3). Periods of economic growth lead to a great demand for shipping, and more ships are built, while periods of stagnation or recession result in a smaller demand. Not always do world trade and shipping move in perfect synchronism, however. In fact, since it can take up to several years for ships to be delivered after the orders have been placed, economic growth may have slowed down by the time the ships are operational. Supply, in other words, responds only slowly to changes in demand. Such conditions can lead to a shortage of ships in periods of strong economic growth, or an oversupply of ships when growth slows down—a situation which accurately describes the current state of the world merchant fleet.

During the late 1960s and early 1970s, Western Europe and Japan went through a period of industrial expansion and strong economic

Table 5.2. **Number of Ships Totally Lost, 1975–1981**

Flag	1975 Number	1975 Gross Tons	1975 Percent Lost	1978 Number	1978 Gross Tons	1978 Percent Lost	1981 Number	1981 Gross Tons	1981 Percent Lost
Canada	5	6,646	0.46	3	2,527	0.13	—	—	—
Cyprus	8	29,873	0.93	31	83,473	3.21	9	11,822	0.65
Denmark	8	50,986	1.14	12	4,004	0.07	8	2,142	0.04
France	2	3,717	0.04	3	12,651	0.10	2	4,641	0.04
West Germany	8	7,199	0.09	6	44,131	0.45	9	20,128	0.26
Greece	18	146,152	0.65	87	782,291	2.30	55	412,107	0.98
India	1	1,348	0.04	7	38,308	0.67	1	9,460	0.16
Italy	1	6,278	0.06	11	17,211	0.15	6	49,137	0.46
Japan	55	66,641	0.17	57	39,011	0.10	29	37,256	0.09
Korea (South)	15	8,431	0.52	17	13,031	0.44	12	30,617	0.60
Liberia	16	249,125	0.38	8	205,550	0.26	7	139,101	0.19
Netherlands	4	1,803	0.03	6	4,145	0.08	4	11,662	0.21
Norway	14	6,219	0.02	17	16,763	0.06	10	2,992	0.01
Panama	42	154,353	1.13	62	223,867	1.08	65	199,354	0.72
Singapore	4	34,348	0.88	6	21,754	0.29	6	24,288	0.35
Soviet Union	3	11,688	0.06	—	—	—	1	5,923	0.03
Spain	23	9,377	0.17	17	8,824	0.11	0	2,846	0.03
United Kingdom	19	62,739	0.19	16	11,447	0.04	9	12,495	0.05
United States	11	16,463	0.11	21	24,394	0.15	11	3,444	0.02

Source. Adapted from Lloyd's Register of Shipping, *Statistical Tables 1982.*

Table 5.3.   **Growth of the World Merchant Fleet;**
1912–1982

Year	Number	Gross Tonnage
1912	23,217	40,518,177
1922	29,255	61,342,952
1932	29,932	68,368,141
1952	31,461	90,180,359
1962	38,661	139,979,813
1972	57,391	268,340,145
1982	75,151	424,741,682

Source. Adapted from Lloyd's Register of Shipping, *Statistical Tables 1982.*

growth, which resulted in a severalfold increase of trade, particularly in natural resources such as coal, iron ore, oil, and phosphate. At the same time, oil imports into the United States increased rapidly and, as a consequence, many new ships, particularly tankers and dry bulk vessels, were ordered throughout this period. The 1973 oil crisis brought a premature end to this period of growth and expansion, but meanwhile the ships, ordered a few years earlier, were launched, causing a massive oversupply (Table 5.4).

This situation persists, particularly in the bulk sector. The tanker fleet was hit hardest. In 1973, oil transports were carried by a fleet of 223

Table 5.4.   **Estimates of Tonnage Oversupply in the 1977 World**
Merchant Fleet

Type of Vessel	Tonnage Oversupply (millions dwt)	Tonnage Oversupply as Percentage of Fleet
Tanker	83.6	25.5
Multipurpose	12.7	26.7
Bulk carrier	26.4	23.4
Residual fleet	18.9	12.3
Total world fleet	141.6	22.1

Source. Adapted from Ramsay, 1980.

dwt. In 1978, when oil transports (in tonmiles) were a mere 10 percent higher, a fleet more than half as large (345 million dwt) was available. It was estimated that in that year tanker surplus averaged some 50 million dwt. As a result, many newly launched tankers went straight from the maiden voyage to a mothballing area. In June 1982, oversupply in the tanker fleet reached a record 56 million dwt, consisting of 349 vessels. Shipowners hope that by the late 1980s the situation will be back in balance.

The dry bulk fleet went through a similar experience. Between 1974 and 1978, transports of coal and grains increased by approximately 10 million metric tons per year, but during that same period, iron ore transports decreased by 65 million metric tons—almost 20 percent. Overall, dry bulk transports (in tonmiles) rose by 10 percent but, as in the tanker sector, the available fleet grew by nearly 50 percent. The number of vessels taken out of operation remained limited, however, because of a decrease in productivity. Since the end of 1979 the size of the dry bulk fleet has not increased and it is expected that the current surplus will disappear by the mid-1980s. The expected increase in the use of coal for power generation will probably require a much larger fleet than is now available, and this may lead to a boom in shipbuilding in the late 1980s.

The conventional cargo fleet did not experience the same type of increase in size but has changed very much in structure since the early 1960s. Conventional cargo is transported by vessels which maintain a regular schedule of sailings between specific ports—the liner fleet—or ships that sail wherever the cargo takes them—the tramp fleet. The share of tramp vessels in conventional cargo transportation is decreasing, while liners have increased their share. There is some overcapacity in the conventional cargo fleet, particularly on certain trade routes such as the North Atlantic, but it is markedly less than in either of the bulk sectors. Projections for the future are made difficult by a variety of political factors, but it is expected that traditional tramp vessels will eventually disappear, whereas the liner sector will continually grow.

A glance at the development of international seaborne trade in the period 1970–1980 (Table 5.5) reveals that the world shipping situation is not improving. In fact, the United Nations Conference on Trade and Development (UNCTAD) secretariat's review of marine transportation for 1981 observed a 2.8 percent decline in the volume of world seaborne trade in 1980, the first decrease since 1975, and this while the world merchant fleet continued to expand slightly—by just under 1 percent

Table 5.5. **Development of International Seaborne Trade, 1970–1980**

| | Tanker Cargo | | Dry Cargo | | | | Total (all goods) | |
| | | | Total | | Main Bulk Commodities | | | |
Year	Millions of Tons	Percentage Increase/ Decrease over Previous Year	Millions of Tons	Percentage Increase/ Decrease over Previous Year	Millions of Tons	Percentage Increase/ Decrease over Previous Year	Millions of Tons	Percentage Increase/ Decrease over Previous year
1970	1,440	13	1,165	13	488	16	2,605	13
1978	1,850	−2.2	1,620	2.7	667	3.4	3,470	0.1
1979	2,003	8.3	1,775	9.6	762	14.2	3,778	8.9
1980	1,806	−9.8	1,886	5.1	796	4.5	3,672	−2.8

Source. After United Nations, *Monthly Bulletin of Statistics*, January 1982; and Fearnley and Egers Chartering Co. Ltd., *World Bulk Trades, 1980*.

Table 5.6. **Structure of the World Merchant Bulk Fleet**

Division of Tonnage	Number	Gross Tonnage
Tankers		
100–499	1,497	519,825
500–999	958	781,992
1,000–1,999	607	913,099
2,000–3,999	575	1,774,250
4,000–5,999	169	802,637
6,000–6,999	54	349,475
7,000–7,999	70	527,896
8,000–9,999	80	726,736
10,000–14,999	454	5,641,068
15,000–19,999	499	8,763,921
20,000–29,999	376	8,968,546
30,000–39,000	338	11,676,995
40,000–49,999	277	12,315,906
50,000–59,999	148	7,976,068
60,000–69,999	127	8,166,626
70,000–79,999	82	6,063,011
80,000–89,999	50	4,211,519
90,000–99,999	31	2,985,001
100,000–109,999	91	9,639,086
110,000–119,999	141	16,337,333
120,000–129,999	158	19,757,082
130,000–139,999	96	13,004,586
140,000 and above	143	24,925,758
Total	7,021	166,828,416
Ore and Bulk Carriers		
6,000–6,999	59	389,117
7,000–7,999	80	600,488
8,000–9,999	331	3,068,644
10,000–14,999	1,220	15,117,346
15,000–19,999	1,287	22,031,582
20,000–29,999	808	19,364,414
30,000–39,999	554	18,991,795
40,000–49,999	160	6,977,713
50,000–59,999	116	6,476,534
60,000–69,999	118	7,673,399
70,000–79,999	108	8,063,630

80,000–89,999	52	4,460,397
90,000–99,999	20	1,885,031
100,000–109,999	3	314,135
110,000–119,999	12	1,394,986
120,000–129,999	9	1,122,463
130,000–139,999	7	938,355
140,000 and above	3	428,024
Total	4,947	119,298,053

Source. Lloyd's Register of Shipping, *Statistical Tables 1982.*

from mid-1980 to mid-1981—further aggravating the tonnage surplus.

In 1980 tanker trades fell by 9.8 percent while dry bulk trades increased by 5.1 percent. Preliminary estimates for 1981 indicate a further decline, resulting from a sharp decrease (13.4 percent) of tanker cargoes and only a small net increase in dry bulk cargoes. Not surprisingly, this decline coincides with a sustained global recession.

Each sector of the world merchant fleet went through tremendous structural changes during the 1960s. In the bulk sector, particularly the tanker fleet, the 1956 closing of the Suez Canal prompted the construction of larger ships, since tankers had to make a 10,000-mile detour to get to the European and American markets. The first 100,000 vessels appeared in the early 1960s and were considered giants. Shortly thereafter, a 150,000-ton vessel was launched, then one of 200,000 tons, and a year later one of 250,000 tons. When a tanker of 326,000 tons was built in 1968, everything seemed possible, though by then the dangers of these colossuses had been demonstrated clearly by the *Torrey Canyon* disaster. Economics of scale also characterized the dry bulk fleet, but certainly not to the same extent (Table 5.6).

The liner fleet was revolutionized by the introduction of containers in the late 1950s. Containers in which the cargo could be stowed were much easier to handle and reduced considerably the time a ship had to spend in port to transfer its cargo. Many special containerships were built and shortly thereafter several other types of intermodal vessels made their appearance, including LASH (Lighter Aboard Ship) vessels which carry barges, and RoRo (Roll-on/Roll-off) ships, enlarged versions of the traditional car ferries.

The liner sector is essentially a self-regulated industry. To prevent predatory and destructive competition, liner companies have organized themselves in shipping conferences which establish a common rate structure followed by all members. The sector operates largely under the principle of the freedom to compete: vessels, regardless of their nationality, can join conferences, or attempt to do so, wherever they expect to operate profitably. As a result, conferences are not only made up of the liner companies from the countries they serve but also of a number of cross-traders—vessels from third countries. This situation has led to the fact that most of the world's liner trade has been carried by the strongly developed liner fleets of the industrial nations (Table 5.7), but there are indications that this will change. The developing nations indeed are seeking a greater share of the world's shipping trade and strongly support the tenet that there is some relationship between the generation of cargo and participation in world shipping. As shown in Table 5.7, developing countries currently generate about 40 percent of the world's cargo but carry less than 10 percent of it. To change this situation in their favor, the fifth United Nations Conference on Trade and Development adopted a Code of Conduct for Liner Conferences, which is based on the concept of the linkage between cargo generation and participation in world shipping. This agreement establishes a 40–40–20 scheme, reserving 40 percent of the cargo for the liner fleets of the countries of cargo origin and destination and no more than 20 percent for cross-traders. The code entered into force on October 6, 1983. Many traditional maritime countries do not like the strict cargo allocation it embodies, but, with the exception of the United States, they made arrangements to adopt it with certain reservations.

Since liner cargoes account for about one fifth of the world's cargoes, the next question is to what extent cargo preference can be secured in the bulk sector, and this is where Third World countries have set their next goal. There is some controversy on whether bulk cargoes, both liquid and dry, lend themselves to cargo allocation, unless there exists a substantial movement between two countries, but the reasons for a cargo-sharing scheme in the bulk trades are ideological, and therefore not always based on common sense. At present, Third World countries export over a third of all dry bulk cargoes and much more of all liquid bulk cargoes, but they carry no more than 9 percent of the total in their ships. The implementation of the Liner Code, followed up perhaps by a similar scheme in the bulk trades, will drastically alter allocation patterns in marine transportation.

Table 5.7. Comparison between Cargo Turnover and Fleet Ownership in 1976

Country Group	Cargo (millions of tons) Loaded	Cargo (millions of tons) Unloaded	Total Cargo (millions of tons)	Tonnage of Merchant Fleet (millions of dwt)	Cargo Turnover (%)	World Merchant Fleet (%)
Developed market economy and flag of convenience countries	1,130	2,544	3,674	521.2	55.4	86.7
Socialist nations	206	133	339	37.0	5.1	6.2
Developing countries	2,038	576	2,615	40.8	39.4	6.8
Total	3,375	3,253	6,627	601.2	100	100

Source. Adapted from UNCTAD, 1978.

5.2. THE REGULATION
OF MARINE TRANSPORTATION

As is the case in the regulation of other marine activities, the 1958 Geneva Conventions codified a number of provisions that applied to marine transportation and thus became international law. The Convention on the Territorial Sea, for instance, provides that merchant ships have a right of innocent passage through the territorial sea of other states. Passage is defined as navigation through the territorial sea for the purpose either of traversing that area without entering internal waters or of making for the high seas from internal waters. It is considered innocent as long as it is not prejudicial to the peace, good order, or security of the coastal state. The coastal state is also given a number of rights: it may require compliance with national regulations in the territorial sea, and it can take the necessary steps to prevent passage that is not innocent. There is still some controversy on the latter subject, however, particularly with respect to the passage of foreign warships.

The Geneva Convention on the High Seas codified the principle of freedom of navigation. As a consequence, the high seas are beyond the control and responsibility of any one state, and it has been necessary to develop an additional body of law to regulate particular aspects of navigation in these waters.

The High Seas Convention also deals with the legal connection between ships and the countries in which they are registered. Generally, the flag state retains exclusive jurisdiction over its vessels on the high seas and, to a lesser extent, when the ship is in the territorial sea of another country. As a result, a ship is never beyond the reach of law since the flag state can exercise jurisdiction over it to prevent unacceptable conduct. The convention further obliges states to ensure compliance with the international navigation regulations. However, as noted earlier, this obligation is not always undertaken by a number of flag of convenience countries.

The rules codified by the 1958 Conventions were in themselves not enough to deal with the many problems of marine transportation. There was, to give one example, a need for measures to prevent collisions or to deal with other safety matters, aspects which the Geneva Conventions left to more specialized organizations, particularly the International Maritime Organization (IMO), to implement.

The principal marine safety agreement is the 1974 International Convention for the Safety of Life at Sea (the SOLAS Convention), which in 1980 superseded a 1960 convention of the same name. Attached to the Convention is a set of regulations on construction, lifesaving appliances, radiocommunication, the safety of navigation, and the carriage of dangerous goods. These regulations are applicable to the ships registered in the countries that ratified the agreement.

IMO also administers a set of traffic regulations aimed at preventing collisions at sea. The first regulations, attached to the 1960 SOLAS Convention, established common standards for lights, signals, and maneuvers. They applied to all vessels on the high seas or in navigable waters, other than ports and certain inland bodies of water. Early on these provisions were perceived to be not entirely satisfactory and the number of accidents, particularly in densely navigated areas, demonstrated that this was indeed the case. As a result, the regulations came under heavy criticism on a number of technical grounds, particularly their failure to incorporate a set of rules on the use of radar and to require compliance with traffic separation schemes, a series of separated traffic lanes IMO had recommended in heavily trafficked areas. Many vessels ignored the traffic separation schemes, sometimes with tragic results. In one classic example, the Peruvian freighter *Paracas* early in 1971 entered the English Channel and, instead of using the northbound lane of the French coast, as she was supposed to do, took the shorter and more convenient lane along the British coast, normally reserved for southbound vessels. The *Paracas* struck the Panamanian tanker *Texaco Caribbean* which exploded and sank with great loss of life. The wreck was marked by British authorities by three vertical green lights, but the story does not end there. The following day, the German freighter *Brandenburg* hit the wreck and sank, taking more than half of her 30-man crew with her. The British then added a lightship and five additional light buoys, but a month later the Greek freighter *Nikki* struck one of the wrecks and went down with her entire crew. Several more buoys and another lightship were added but shortly thereafter the American aircraft carrier USS *Enterprise*, ignoring the flashing lights from the lightship, ran through the buoys but, fortunately and to everyone's surprise, got away with it.

To prevent such occurrences, IMO in 1972 convened a diplomatic conference in London which produced the Convention on the International Regulations for Preventing Collisions at Sea, also called the Collision Regulations or simply COLREG. This agreement went into force in July 1977, making the use of traffic separation schemes compulsory. This does not mean that everyone is obediently following the schemes. Violations, in fact, occur every day but, as is the case in practically the entire body of international navigation law, only the flag state has jurisdiction over them, unless the violation took place within the jurisdiction of another state.

A great number of accidents are caused by human error. Several studies have indicated that most collisions are attributable to poor seamanship and could have been avoided if prudence and alertness had been shown. Moreover, these investigations have pointed out, as should be of no surprise, that numerous mistakes appear to be made aboard flag of convenience vessels. Error can never be ruled out as a factor, but if errors are caused by a lack of training or poor seamanship, their chance of occurring can be significantly reduced by stringent regulations.

The International Labor Organization has concerned itself with this subject and recently adopted a Convention on Minimum Standards in Merchant Ships along with a number of recommendations to its already extensive *International Seafarer's Code.*

Following a number of accidents which demonstrated the appalling lack of training and the number of uncertified people manning the bridges of flag of convenience vessels, IMO entered the picture with the adoption of the 1978 Convention on Standards of Training, Certification and Watchkeeping, which entered into force on April 28, 1984.

Another important agreement on maritime safety promoted by IMO is the 1966 International Convention on Load Lines, which provides for the painting of Plimsoll lines on ships. Moreover, the convention requires contracting governments to prevent ships from putting to sea without the appropriate International Load Line certificates, and this essentially to avoid overloading, which increases the stress on a vessel significantly.

Finally, a number of treaties deal with some of the jurisdictional problems that emerge with collisions and other incidents. The International Convention for the Unification of Certain Rules Relating to Penal Jurisdiction in Matters of Collision or Other Incidents of Navigation provides that, in the event of a collision involving the responsibility of someone on the ship, only the flag state may initiate criminal proceedings or arrest the ship. The International Convention Relating to the Arrest of Sea-going Ships, on the other hand, established the principle that a ship from a contracting state may be arrested within the jurisdiction of another state in respect of any maritime claim. Finally, the International Convention on Certain Rules Concerning Jurisdiction in Matters of Collision seeks to unify the rules of private international law by giving plaintiffs the right to sue before a court in the country where the defendant lives, where the collision took place, or where the arrest was made.

In spite of these agreements, all of which were concluded in the 1950s, it still could be very difficult to find someone responsible for a mishap. The classic example was provided by the *Torrey Canyon*, which was owned by an American company that had leased the vessel to a British company. The ship, built in the United States and rebuilt in Japan, was registered in Liberia, insured in London, and manned by Italians. The United Kingdom and France, which suffered most of the damage caused by its grounding in 1967, did get their money, but not through the conventional channels. Instead, they pretended they were not looking, but when one of the *Torrey Canyon*'s sister ships entered Singapore (part of the British Commonwealth), they arrested and held her until the insurers paid a $7,500,000 damage settlement. Fortunately, the situation has improved somewhat with the adoption of the various liability agreements enacted since the *Torrey Canyon* incident (see Table 4.9), though the immense damage cause by disasters of *Amoco Cadiz'* magnitude adds entirely new dimensions to the problem.

Chapter Six

Energy from the Sea

Since conventional energy sources will not last forever, and nuclear energy, though virtually inexhaustible, raises serious environmental and safety questions, the search for alternative energy sources becomes more important. One of these alternatives is the sea.

6.1.   OVERVIEW

Covering more than 70 percent of this planet's surface, the oceans indeed represent the largest solar energy collector available. Though the majority of the sun's rays are bounced back, a substantial portion of the energy is converted and stored, and it can later be extracted.

Thermal energy, for instance, is stored as a temperature difference between huge volumes of warm surface water and cold deep water. This difference can be used to drive heat engines, a process known as ocean thermal energy conversion—OTEC in short. Mechanical energy is stored by waves and currents. It can be extracted from waves by using mechanical devices connected to electric generators and from currents by using the equivalent of underwater windmills. Finally, physical energy is stored by the action of sunlight establishing a different salt concentration. This can be exploited using membranes similar to those used by desalinization plants to develop large hydraulic pressures, which can be used to drive turbines.

Of course; not all of these sources hold the same potential, first because they do not have equal power levels, and second because some are easier to extract than others. Many ideas have been suggested to tap each potential source, however. Though most of these will never become operational, some will, helping to alleviate the world's energy needs in the coming decades.

6.2.   SOURCES AND METHODS

OTEC

Among the various sources of marine solar energy, OTEC appears quite a bit farther along the road to commercial development than the others. As previously noted, the system relies on the natural temperature differences between the warm surface and the cold deep waters of the sea. Surface waters, heated by the sun, easily reach more than 25°C in tropical areas, whereas deeper water, cooled by polar currents, has a temperature of barely a few degrees. The difference, usually around 20°C, is enough to produce electricity.

There are two methods to do this. The first, called the closed cycle system, uses a rapidly boiling work fluid such as ammonia (Fig. 6.1).

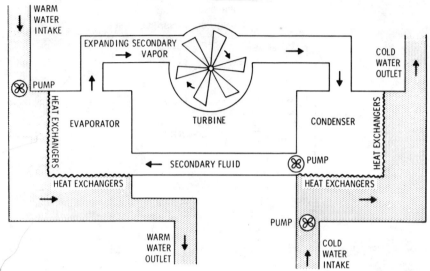

Figure 6.1.   Diagram of the OTEC closed-cycle system. The secondary fluid is a rapidly boiling fluid such as ammonia, which is evaporated by the warm surface water and condensed by the cold deep water.

Warm water is drawn in from the surface and brought in contact with the working fluid via extemely sensitive heat exchangers. The ammonia evaporates and the steam can be used to drive the turbine of a conventional generator. Cold water, pumped up from great depths, is then used to condense the gas to its liquid state and the cycle can start all over. The second method, known as the open cycle system (Fig. 6.2), draws the warm water into a flash evaporator, where a series of four chambers successively lowers the pressure of the seawater to a level below its saturation point to cause a flashing off of steam. The vapor is forced through a turbo generator to produce electricity and condensed by cold water so that it can be recuperated as fresh water or returned to the sea. In both instances, no fossil fuels are used during normal operations; the ocean is the energy source.

Ocean thermal energy conversion is not a new development. It was, in fact, already proposed in the early 1880s by the French physicist Arsène d'Arsonvalle. A small plant, following the open-cycle system, was built in Cuba in 1930 by the French inventor Georges Claude, but it lasted no more than a few weeks before being destroyed by heavy seas. During the next 30 or 40 years the idea languished until it was

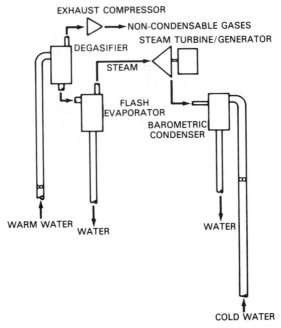

Figure 6.2.   Diagram of the open-cycle OTEC system, which uses flash evaporation of seawater under a partial vacuum.

taken up again in the mid-1960s and rapidly gained support throughout the oil-crisis–marked 1970s.

In those few years, OTEC technology has come a long way. Much of the analytical work and many preliminary designs have indeed been verified by various programs, and the at-sea operational capacity of the system has been demonstrated. One of the most important test programs was MINI OTEC, which became the first at-sea OTEC plant to produce power. The program was developed by a private consortium, including the University of Hawaii and several companies, and operated off the Hawaiian coast at Keahole Point. To pump up the cold water, a 630-m-long polyethylene cold water pipe, 0.6 m (about 2 ft) in diameter, was deployed. However small the electrical output (50 kW), MINI OTEC demonstrated that the system worked and provided a great amount of data on biofouling and heat exchangers. The plant suspended operations in December 1979.

A year later, a 1 megawatt (MW) test facility, called OTEC 1 (Fig. 6.3) was evaluated. Using a mothballed T-2 U.S. Navy tanker, the U.S. Department of Energy tested heat exchangers and biofouling countermeasures. Despite the short period of operations—no more than four months as a result of a cut in funding—most of the tests were concluded and OTEC 1 performed to all expectations. Another major at-sea test, in support of a 40-MW pilot plant, involved a ⅓ scale test of a fiber-reinforced plastic cold water pipe, 3 m in diameter and 300 m long. The major thrust of present research efforts is to initiate the preliminary design for a 40-MW pilot plant, with subsequent construction and deployment by the late 1980s. Even if this occurs as planned, the larger (100–400 MW) commercial-size OTEC plants should not be expected in the near future. Many technological problems remain to be solved, particularly those concerning platform design, the mooring system, the seawater system, the cold water pipe, and the power transmission cable.

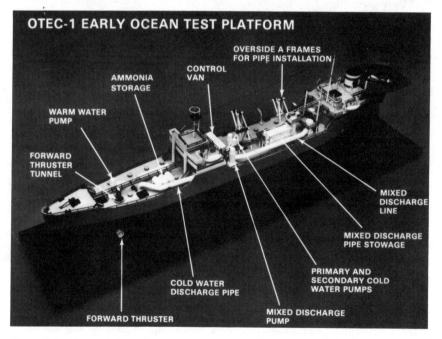

OTEC-1 EARLY OCEAN TEST PLATFORM

OVERSIDE A FRAMES FOR PIPE INSTALLATION

CONTROL VAN

AMMONIA STORAGE

WARM WATER PUMP

FORWARD THRUSTER TUNNEL

MIXED DISCHARGE LINE

MIXED DISCHARGE PIPE STOWAGE

PRIMARY AND SECONDARY COLD WATER PUMPS

COLD WATER DISCHARGE PIPE

MIXED DISCHARGE PUMP

FORWARD THRUSTER

Figure 6.3.   Some of the major power components of the OTEC-1 engineering test facility.

Platform Design

The selection of the platform for an OTEC installation is based on such factors as motion response, the extent of forces induced in the cold water pipe, the ability to effectively house and support the heat exchangers, energy conversion equipment and auxiliary equipment for closed-cycle operations, and, of course, construction costs. Several configurations for a commercial floating OTEC plant have been investigated.

One design involves a plant moored in very deep water close to the coast, transmitting electrical power to shore via a power transmission cable (Fig. 6.4). Another concept, the so-called grazing plant, operates as a self-contained plant ship on which an energy-intensive product, such as ammonia or hydrogen, is produced. Its main advantage is that the plant can cruise (or "graze") around tropical waters to track the largest available temperature differences. A number of designs are very similar to the bottom-fixed structures used around the world by the offshore petroleum industry. The key advantage of such structures is that they eliminate the high technical risks associated with the dynamics of the cold water pipe, the mooring system, and the power transmission cable. Finally, concepts for land-based OTEC facilities have been studied as well. They are attractive in that they eliminate many ocean engineering problems, but they would require unique sites in tropical areas on islands with a readily accessible cold water supply and a sufficient energy demand.

Commercial studies have indicated that the larger plants are more cost effective in terms of product efficiency versus size. They also minimize platform motions and consequent stresses induced to the cold water pipe, but there are limitations, particularly concerning their constructibility, transportation, and deployment. These considerations would limit, at least initially, the size of commercial plants from 100 to 400 MW, which are still enormous structures. A 400-MW plant, for instance, would require a 560,000-ton surface ship. As this is comparable in size to some of the platforms currently operating in the North Sea, it is safe to assume that commercial-size OTEC plants can be built with existing technology, though there are still many unknowns in terms of the total integration of the OTEC components with the platform. A problem requiring resolution prior to commercialization, for instance, is the development of a reliable, long-life coupling to connect and disconnect the pipe from the platform. The dynamic action of the platform with the cold water pipe also remains to be validated. Taking account that all this must be deployed and operated

at sea over a design life of some 30 years, it is clear that there is quite
a bit of work yet to be done.

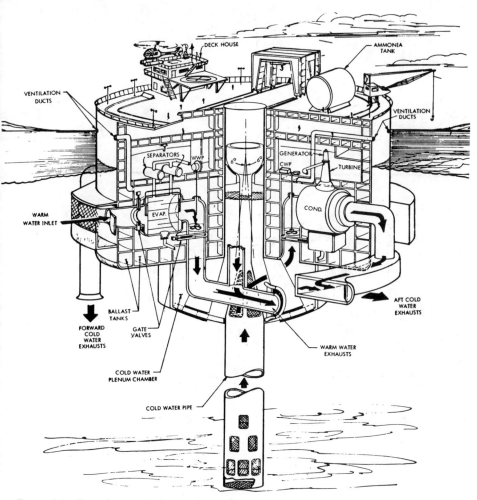

Figure 6.4.   Two designs for large-scale commercial OTEC plants. The first, developed
by the TRW Systems Group, Inc., includes four 25-MW power modules (one of which is
shown in the cutaway portion), which could generate a total of 100 MW. The platform
shown would have a diameter of 100 m. The second concept, developed by the Lockheed
Missiles and Space Company, has four 65-MW power modules, for a total production of
260 MW. The cold water pipe, shown on the right side of the diagram, would have a
diameter of 38 m. Also shown in this section is the mooring system, the anchor, and the
power transmission cable. (Courtesy TRW Systems Group, Inc. and Lockheed Missiles &
Space Co., Inc.)

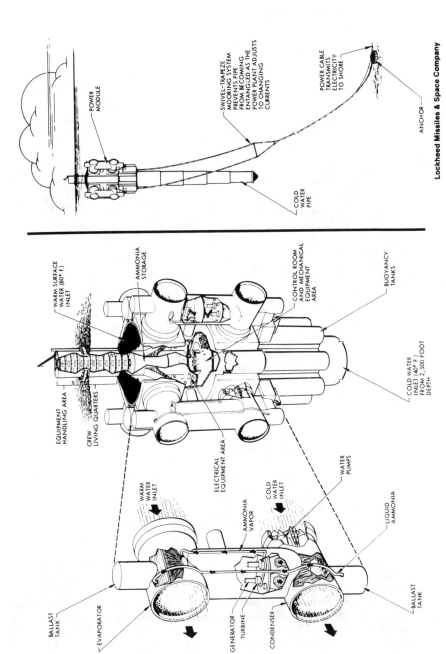

POWER MODULE

SWIVEL-TRAPEZE MOORING SYSTEM PREVENTS PIPE FROM BECOMING ENTANGLED AS THE POWER PLANT ADJUSTS TO CHANGING CURRENTS

POWER CABLE TRANSMITS ELECTRICITY TO SHORE

ANCHOR

COLD WATER PIPE

Lockheed Missiles & Space Company

WARM SURFACE WATER (80° F)
AMMONIA STORAGE
CONTROL ROOM AND MECHANICAL EQUIPMENT AREA
BUOYANCY TANKS
EQUIPMENT HANDLING AREA
CREW LIVING QUARTERS
ELECTRICAL EQUIPMENT AREA
COLD WATER INLET (40° F) FROM 2,500 FOOT DEPTH

BALLAST TANK
EVAPORATOR
WARM WATER INLET
AMMONIA VAPOR
GENERATOR
TURBINE
CONDENSER
COLD WATER INLET
WATER PUMPS
LIQUID AMMONIA
BALLAST TANK

Figure 6.4. (continued).

Mooring System

Offshore OTEC plants require deep water mooring systems to limit the movement of the platform and to minimize flexures in the power transmission cable. Though there are deep water mooring systems in existence, little experience is available on their prolonged life requirements. Hence most of the remaining problems in designing a mooring system for a 30-year expected plant life relate to unknowns in fatigue and corrosion aspects.

There is also a need for a better method of predicting wave drift forces and the dynamic response of the total OTEC plant. Seafloor engineering investigations at the planned sites must be made as well to obtain data on soil properties and seafloor stability.

Seawater System and Cold Water Pipe

The array of warm and cold water inlets and outlets, pumps, pipes, condensors, and evaporators is usually referred to as the seawater system (SWS). It is with these components that the least amount of experience exists.

The OTEC seawater pumps, for example, present very challenging design problems since they have to take care of unprecedentedly high-volume flow rates and must be able to withstand seaway motions of the platform, which could conceivably cause significant pressure differences. The pumping rates are simply staggering. The cold water pumping requirements of a 400-MW plant, for instance, amount to about 3000 $m^3/$s. With an equivalant rate of warm water, inlets and outlets handle a flow rate comparable to that of the Nile!

The heat exchangers have to be extremely sensitive and efficient, as well as resistant to corrosion. Assuming a temperature difference of 20°C, the maximum energy recovery is just over 30 percent. In a real system, however, an efficiency of 2.5–3.5 percent seems more likely, because losses inevitably occur. A conventional coal-powered generating plant, in comparison, can be as efficient as 30 percent. For an OTEC plant to be as efficient it must, in other words, transfer about 10 times as much heat to yield a similar output. This is still possible because of the enormous quantities of warm water that are available.

The cold water pipe (CWP) is another critical component of the system, mainly because of its size and immense moment of inertia when handled during construction, transportation, and deployment. One design for a 400-MW plant indeed uses a single cold water pipe, 900 m long and 30 m in diameter. Several smaller pipes could also be used, but even so it remains quite a structure to deploy in a largely hostile environment. To complicate matters, account must be taken of biofouling, corrosion, and loads on the system, which are still major areas of uncertainty.

Power Transmission Cable

Also in what concerns the power transmission cable, which will transport the electrical power from the OTEC plant to shore, a lot of unknowns need to be verified. This is particularly the case with respect to the riser cable, extending from the platform to the ocean floor, which will be subject to mechanical loads outside the long-term experience base of the industry. For bottom-laid cables at OTEC power levels, cable design methods and performance standards exist.

Distribution of OTEC

Since a temperature difference of at least 20°C is needed to make an OTEC plant work effectively, the distribution of the OTEC resource is limited to tropical regions, approximately between the tropics of Cancer and Capricorn (Fig. 6.5). Other limitations of the distribution of plants include the proximity of a market and the oceanic environmental impact, that is, the whole gamut of chemical, biological, and physical factors that influence its operation and may be different from one region to another. Particularly promising candidates for OTEC plants are island communities in tropical regions, though they often would better be served by a smaller plant (40–100 MW), unless they could market surplus power. In various proposals, mention is often made of such places as Puerto Rico, Hawaii, Guam, and the Virgin Islands. Possibilities for plants in the 400-MW size exist off the coast of Florida or at any site off the U.S. coast with a sufficient temperature difference within a distance of not more than 100 miles.

Until recently, OTEC research was conducted primarily in the United States. To foster OTEC's development, the U.S. Congress enacted two laws in 1980. The first established a technical review panel to review OTEC and the progress made on its development. Moreover, Congress was authorized to appropriate up to $20 million to "provide a technological base for 100 MW of OTEC generated electricity by 1986, 500 MW by 1989 and 10,000 MW by the year 2000." These funds were not intended for actual construction costs but for the development of plans leading to these goals. The second act gave OTEC legal, corporate, and governmental identity. It determined a number of regulations for OTEC operations and also set up an "OTEC Demonstration Fund" for the construction of demonstration plants. While all this seemed very promising at the time, ocean energy funds were rather drastically cut by the Reagan Administration, and this will inevitably lead to delays in the OTEC program.

Japan is also heavily involved in OTEC research. As an island nation, it realizes OTEC's potential and recently increased the OTEC budget by some 60 percent. An experimental 1-MW OTEC plant was tested off Kumejima in Okinawa. Next in line are plans for a 10-MW pilot plant and a 100-MW demonstration facility.

In Europe, both Sweden and the Netherlands are actively involved in OTEC development. Dutch research and development efforts have focused on the cold water pipe, the floating platform, and the mooring system. In January 1980, the Dutch government decided to fund technical and economic studies for a 10-MW demonstration unit. In Sweden a group was formed in 1979 as a follow-up to the cooperative work initiated within EUROCEAN's OTEC working group. This organization plans to design and build complete plants for export purposes. Initially, this will involve 10-MW units, incorporating desalinization capacities where appropriate.

Finally, it is worth mentioning that environmentally the OTEC system does not seem to be much of a problem. Of course, if the working fluid were to be released in the sea, this could create some nasty problems, but in general it appears that the environment will do more damage to OTEC than OTEC will do to the environment.

Waves

Waves, which are in fact degraded solar power created by winds, contain a substantial amount of energy. There are three principal methods to

ΔT(°C) BETWEEN SURFACE AND 1000 METER DEPTH

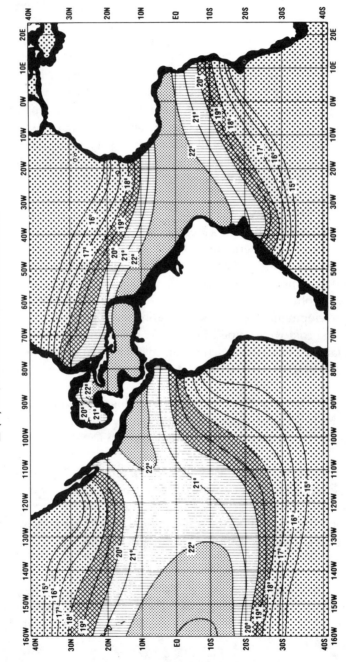

- AVERAGE OF MONTHLY ΔT's LESS THAN 18°C
- AVERAGE OF MONTHLY ΔT's MORE THAN 18°C, LESS THAN 20°C
- AVERAGE OF MONTHLY ΔT's MORE THAN 20°C, LESS THAN 22°C
- AVERAGE OF MONTHLY ΔT's MORE THAN 22°C, LESS THAN 24°C
- AVERAGE OF MONTHLY ΔT's GREATER THAN 24°C
- WATER DEPTH LESS THAN 1000 METERS

136

ΔT(°C) BETWEEN SURFACE AND 1000 METER DEPTH

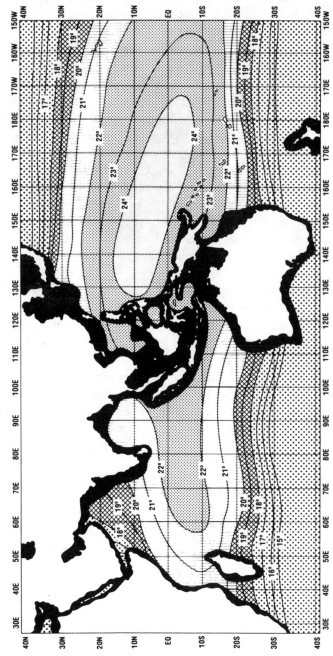

Figure 6.5. Worldwide distribution for the ocean thermal resource. Contours indicate the annual averages of monthly temperature differences in degrees Celsius between the ocean surface and depths of 1000 m. Areas with temperature differences greater than 20°C are suitable for OTEC development. The black areas in coastal regions denote water depth less than 1000 m.

extract this energy: techniques that use the vertical rise and fall of successive waves to drive a turbine; techniques that use the rolling motion of the waves to move vanes or cams; and techniques that converge waves into channels to concentrate their energy.

The first method, vertical displacement, is perhaps the easiest entrée into wave power. It has been used for several years to produce small amounts of energy, particularly in navigational aids such as lighted or whistler buoys. A larger scale power plant is being built on this principle in Japan. It uses the up and down motion of waves to trap air, which in turn drives a turbine. One of the designs to generate electricity from waves (Fig. 6.6) consists of a vertical pipe containing a one-way check valve and a buoyant float at the surface. When the float and the pipe descend into a wave trough, the water flows up the pipe past the check valve. As the float ascends a wave crest, the water is prevented from flowing down by the check valve. Subsequent cycles elevate the water into a reservoir until pressures are reached suitable for power generation by the flow of water through a turbine. Systems of this type will not generate major amounts of power.

Another design involves a pump consisting of a vertical riser with a flapper valve and a buoyant float at the surface. It is loosely attached to the bottom and can move directly with the motion of the waves. When in operation, the valve is closed for about half of the wave cycle, allowing the water in the column to move upward with the float into a reservoir. Through the continuous movement of the pipe, the hydraulic pressure in the reservoir increases, so that the water can be allowed to escape under pressure to turn a turbine. A series of such wave water pumps could be used to provide a modest supply of electricity, which could be useful near offshore drilling sites or similar structures where a power source is needed.

The second method of power generation from waves, using the rolling motion of waves, is not yet well developed. Two designs have been given serious thought, however. One, in which the British have invested considerable research and development, is called a Salter's Duck. It consists of a series of segmented cam lobes that are randomly rocked on a large spine by the waves. Pumps connected to the lobes send high-pressure water through small pipes to drive a generator. In principle, in a regular idealized wave train, this system can use wave energy with considerable efficiency, but it remains to be seen whether the Salter's Duck is practical when exposed to real waves on an open coast.

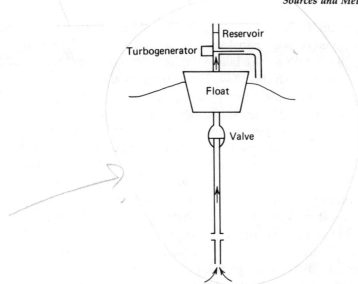

Figure 6.6.   The vertical displacement method. (Adapted from Isaacs and Seymour, University of California, Institute of Marine Resources, Refs. 74-79, La Jolla, California, 1974.)

Another concept, the hydroelectric system, converts wave energy to electricity through a series of rocking floats connected to piston pumps. The high-pressure seawater carries the energy to a central powerhouse, where it feeds a variable ratio hydraulic transformer. The output of the transformer is a closed freshwater system operating at constant pressure, feeding a Pelton wheel type turbine running at constant RPM. The electrical output of the alternator can then be fed ashore or to floating electrochemical plants.

Among the appealing aspects of using waves as a source of energy is that these devices can be used almost anywhere and are nonpolluting. They can cause environmental problems, however, if they interfere with the breaking waves or the movement of sand in the coastal zone. On the other hand, if properly placed, this impact can be used to absorb energy from destructive coastline waves and protect beaches and property from coastal erosion.

Currents

Ocean currents such as the Gulf Stream and the Kuroshio may also become a source of energy in the future. Even though they have very low energy densities, the kinetic energy in major currents is impressive.

The Gulf Stream, for instance, channels 30×10^6 m³ of water per second at a surface speed of more than 4 km/h. According to the U.S. Department of Energy, 10,000 MW or 87.6 billion annual kilowatt hours theoretically could be extracted from this flow. The world's total current resource is estimated by Scripps Institution of Oceanography scientists at 5×10^6 MW or nearly 44×10^{12} annual kilowatt hours.

The devices that have been suggested to tap energy from ocean currents are in principle rather simple (Fig. 6.7). They include a tethered free-stream, four-stage, six-bladed underwater mill; a tethered free-stream Savonius rotor power station; and a tethered, four-disc, axial-flow water turbine. Another concept consists of a series of parachutes attached to a continuous cable. All these systems have in common that they would be difficult to build and maintain. In fact, it would almost seem necessary to intensify the current by means of a structure to increase water velocities and reduce the dimensions of the turbine—much the same way the gradient of rivers is changed by dams to achieve higher velocities or pressures over a smaller spatial interval. Such structures would again be difficult to build and maintain, so that the extraction of energy from ocean currents remains an option only for the distant future.

Figure 6.7.   Artist's rendition of *Coriolis One*, a huge underwater current turbine, being towed to its mooring. (From K. Stowe, *Ocean Sciences*, 2nd ed., Wiley, New York, 1983.)

Tides

Diurnal waves, or tides, also contain energy, but so far it has not been extracted effectively. Two plants are currently in operation: one on the Rance in France, another in Kislaya Guba on the Ura in the Soviet Union. The Rance plant is a 750-m-long structure damming a reservoir of 184×10^6 m^3. Tides with an amplitude of up to 13.5 m drive submerged turbines during both incoming and outgoing tides, to produce about 240 MW. The Russian plant is much smaller.

Although there are problems with tidal systems—for instance, tides do not flow continuously and vary in strength—the price of fossil fuels may make tidal power more attractive. Areas where tidal power plants could be installed include the coasts of Alaska and British Columbia, the Gulf of California, the Bay of Biscay, the White Sea, the central Indian Ocean, and the coasts of eastern Maine and Canada.

The construction of tidal plants does not require any major technological breakthroughs. Other advantages include a longer life than conventional energy plants and, of course, tides are free—though a tidal difference of less than 5 m would probably remain impractical. For the moment, however, this source of energy is not competitive with conventional energy sources.

Salinity Gradients

An interesting but still uneconomical method of producing power from the sea would be to utilize the osmotic or salinity gradient power resulting from the osmotic pressure difference between freshwater and saltwater. This pressure difference is indeed equivalent to a head of water of 240 m. In terms of power, 1 m^3 of freshwater dissolved in a large amount of seawater will dissipate about 2.25 MW. Of course, not all of this is recoverable, but if the world's total river supply is considered, a theoretical supply of 2×10^{12} W is possible—more than the present world consumption of electricity. Though significant in size, it will take extremely complex technology to tap this energy source.

Several approaches have been suggested for extracting power from salinity gradients. These include a vapor exchange process between two solutions, called inverse vapor compression; simple osmotic exchange against a hydrostatic pressure, referred to as pressure-retarded osmosis; and the dialytic battery, which can be thought of as inverse electrodialysis.

Only the first of these does not require membranes, with their attendant expense, relatively short life, polarization problems, and high pumping costs. The second method is a two-step process in which the river water falls a distance below sea level, whereafter the freshwater is driven into the sea through semipermeable membranes. The dialytic battery, on the other hand, is composed of a series of alternating anion and cation exchange membranes forming separate compartments, through which waters with different salt concentrations can flow, with a power source at each end of the stack. The flow of elements across the membrane will then form an electrical charge between the electrodes. Such a battery is technically feasible but not economically practical.

Although salinity gradient energy is a very large potential source of power, ranking in both total magnitude and energy density well above the other marine energy options, it is not certain whether it can be commercially used. Moreover, it is clear that—if implemented—these extraction methods would have considerable environmental effects.

Biomass Conversion

A final method to extract energy from the sea is by converting photo-synthetically produced organic matter—or biomass—into fuel. The oceans indeed fix some 10^{10} or more tons of carbon per year into organic material, and some of it rather efficiently in large marine plants such as kelp. The biomass conversion method relies on the growing of such plants for anaerobic decomposition with the ultimate production of gas. The gases produced, usually a mixture of carbon dioxide and methane, have heating values of 500–800 Btus per standard cubic foot and can be readily upgraded to pipeline quality gas by established procedures.

The difficulty with this approach lies in the fact that it takes a lot of biomass to make a significant contribution to the energy budget. It has been estimated, for instance, that, using the present yields of U.S. crop-lands, it would take some 88 million ha of cropland converted to methane by the process of anaerobic digestion to produce 1 quad (10^{15} Btus) of energy—about 1 percent of the projected U.S. energy budget for the year 2000.

Still, since a lot of the ocean is uncultivated and since some marine plants grow extremely fast, it seems possible that marine biomass, some time in the future, could be grown for energy conversion. For that purpose, kelp farms have been designed and some tests have been undertaken.

Economic analyses, however, have cast considerable doubt on cost-effectiveness and the energy input–output ratio of such farms. It will take a lot more research on the basic biology of the plants, particularly with respect to their nutrition and growth, before adequate and reliable yields can be produced. When, and if, this is done, marine biomass can make a contribution to liquid fuel supplies that can be used as a substitute for petroleum products.

Chapter Seven

The Law of the Sea

I nternational law can be broadly defined as that body of rules which sovereign nations have agreed to observe in their relationship with each other. Its main purpose is to establish an order based on justice and peaceful relations among states. There is, however, no legislature to lay down these rules. Even organizations such as the United Nations are not designed to adopt rules binding on those states which do not choose to accept them. This is a consequence of the principle that states, no matter how great or small, are equal and sovereign, and that no state can legally impose its will on others. In practice, of course, this is not always the case.

There are two ways in which rules of international law can be developed. States may be bound by rules of customary international law. When the practice of states on a particular matter achieves a certain degree of consistency, the international community may come to feel bound to observe this practice, and it is then considered to have crystallized into a rule of customary international law. States may also enter into legal obligations by treaty. These obligations are binding on all the states which become party to the treaty, but not on the states that do not ratify it.

The law of the sea is one of the oldest components of international law. From the moment we started using the oceans, a number of practices and principles developed, some of which gradually became accepted as customary international law. Many of these principles were codified by the four 1958 Geneva Conventions on the Law of the Sea, often referred to throughout this book, which form the basis for contemporary ocean law. The Geneva Conventions merely provided a framework, however, and specific ocean activities became regulated by a number of additional treaties.

Given the importance of the law of the sea in the management of marine resources, this chapter reviews the present regime and what led up to it. In addition, the future regime, as included in the new Law of the Sea Treaty, is briefly discussed.

7.1. THE LAW OF THE SEA BEFORE 1958

From a Western perspective, the law of the sea originated in and around the ancient Mediterranean. The Mediterranean at this early time was crisscrossed by many trade routes and, since marine transportation could involve nationals from several states, a body of law gradually developed to deal with potential conflicts and disagreements or other matters relating to the conduct of navigation. This early set of ocean law was practiced, refined, and amended on the Greek island of Rhodes, a very important trading center, and became known as Rhodian sea law. Though it dealt primarily with commercial relations between private citizens or entities of various states (private international law), Rhodian sea law did include some concepts of importance to the interaction between their governments (public international law). The most important of these stipulated that the seas were free to be navigated, or used for that matter, by all nations, a concept which would gradually develop into the doctrine of the freedom of the seas. Even so, though neither the Roman Empire nor the Greek city-states were true maritime powers, they exercised various types of jurisdiction over their adjacent waters.

This practice was followed by several Italian city states during the early Middle Ages. None of these early practices resulted in any clear and recognized claims to sovereignty over the sea until Venice, through sheer power, managed to enforce its claim to jurisdiction over the entire Adriatic Sea during the fourteenth century.

From then on, two philosophies on the law of the sea developed. One was based on the freedom of the seas, the other on the right of states to claim sovereignty over their adjacent waters. The latter practice was particularly adhered to in the Mediterranean, but attempts also were made by Scandinavian states to control fishing and navigation in waters as far away as Greenland. Nevertheless the most extensive claim was made by Spain and Portugal, which, through a series of papal bulls and the Treaty of Tordesillas in 1494, went as far as dividing up all of the oceans in the Southern Hemisphere.

Throughout the sixteenth century, the debate between those countries favoring the freedom of the seas (*mare liberum*) and those advocating their right to claim extensive areas (*mare clausum*) grew stronger. England, under the nationalistic Stuarts, left the group of nations favoring the freedom of the seas and began to claim the waters surrounding the British Isles. This led to conflicts, particularly with the Dutch, who possessed an impressive merchant fleet and operated an extensive fishing fleet in British waters. To justify the Dutch reaction, Holland's leading lawyer, Hugo Grotius, early in the seventeenth century spelled out the arguments in favor of the freedom of the seas. His philosophy asserted that, since the waters of the sea are not susceptible to effective occupation and the resources of the sea are inexhaustible, the oceans should be free and open to all. England responded through one of its leading lawyers, John Seldon, who argued in favor of a closed sea, but it was the Grotius doctrine that ultimately prevailed. After achieving naval supremacy during the eighteenth century, England could afford to be more moderate, and the doctrine of *mare clausum* gradually disappeared from practice.

While national claims to extensive ocean areas failed to win acceptance in international law and practice, the concept of a narrower belt of waters along the coast, over which a state had complete sovereignty, appeared to have more appeal. At various times and places in Europe, territorial waters were considered valuable and useful for protecting coastal fisheries, but there was no agreement as to whether they should include all water in sight of land, all water which could be defended by a shore-based cannon, or whether they should be based on some other criterion.

Cornelius van Bynkershoek, a Dutch judge, finally popularized the cannonshot rule, which made sense because this was about as far out to sea as a state could appear to have both command and possession. Other writers and states followed his suggestion with increasing frequency, and gradually a trend emerged to equal the distance of a cannon shot with one marine league (three nautical miles), though it is doubtful that cannons shot that far at the time.

The three-mile distance, however, was never accepted as universal international law. The Scandinavian countries claimed four miles, several Mediterranean countries held to six miles, and others asserted jurisdiction over even larger zones or made no claims at all. In 1930, the League of Nations attempted to create some uniformity by including a number of ocean law issues on the agenda of the Conference for the Codification of International Law at The Hague. Such matters as a three-mile territorial sea and exclusive fishery zones were discussed, but the Conference was unsuccessful in its attempt to codify the law of the sea. Even so, it performed a most useful function by identifying and partially defining many issues that were to grow steadily in importance.

The first major signal in the movement toward a new ocean order came with the Truman Proclamations of 1945, which asserted U.S. control over the resources of its continental shelf and the management of its fisheries. By doing so, the United States called the world's attention to the notion that there was something of great value besides fish in the sea, and that nothing in international law prevented a coastal state from claiming it. Soon other countries were following this example. Only a month after the Truman Proclamations, Mexico claimed similar rights in a proclamation signed by President Avila Camacho. A year later, Argentina claimed not only the resources of its extraordinarily broad continental shelf but also those of the superjacent waters. Peru and Chile acted similarly by asserting a 200-mile zone in 1947, thus claiming the extremely rich fisheries of the Peru current. Within a few years, maritime claims multiplied and escalated.

While this rush for extensive ocean areas was gathering momentum, the United Nations was embarking on its task of developing and codifying international law. The law of the sea was given high priority, and to prevent conflicts or confusion the General Assembly asked the International Law Commission, a United Nations body of judicial experts, to prepare draft articles concerning the high seas, the territorial sea, the continental shelf, and the regime of fisheries. After seven years of work, the Commission produced four draft conventions, which were discussed at the first United Nations Conference on the Law of the Sea, convened in Geneva from February to April 1958.

7.2. THE 1958 GENEVA CONFERENCES ON THE LAW OF THE SEA

The 86 states that participated in the first United Nations Conference on the Law of the Sea, impelled by a sense of historical necessity, adopted four conventions: the Convention on the Territorial Sea and the Contiguous Zone, the Convention on the High Seas, the Convention on the Continental Shelf, and the Convention on Fishing and the Conservation of the High Seas. They were ratified by a sufficient number of states to enter into force within a few years. Much of the traditional law of the sea, which had developed as customary law over the centuries, and a number of new concepts such as the continental shelf, were thus codified in international treaty law.

The Geneva Conventions divided the oceans into different zones. Coastal states could exercise various degrees of jurisdiction over the internal waters, the territorial sea, the contiguous zone and continental shelf; the other zone was the high seas, which remained a common zone where the principles of the freedom of the seas applied (Fig. 7.1). The

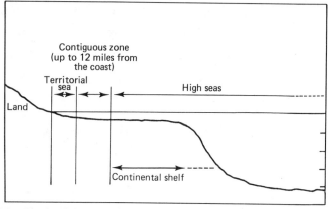

Figure 7.1. The legal division of the ocean following the 1958 Conference on the Law of the Sea.

Conference failed, however, to resolve some of the most difficult controversies, including the establishment of a uniform rule on the breadth of the territorial sea and the exclusive fishing zones. In addition, the Conventions included a number of ambiguous definitions and procedures, some of which might have been acceptable at the time but required clarification soon thereafter.

To resolve these problems, a second Conference on the Law of the Sea was convened in Geneva in 1960. This conference came very close to a solution on the territorial sea. A United States–Canadian proposal which envisaged a territorial sea of six miles combined with a fishery zone of an additional six miles indeed seemed to carry wide support but, when coming to a vote, narrowly failed to get the necessary majority. The Geneva Conventions, as a consequence, remained unchanged, though it was clear they were becoming rapidly outdated.

7.3. PREPARING FOR A NEW OCEAN ORDER: 1960–1973

New economic and political conditions after 1960 gave rise to changes in the concepts of national sovereignty and economic rights, and to increasing demands for a revision of the existing ocean regime.

There were many reasons for this. The four Geneva Conventions, far from being universally accepted, left a number of serious matters unsettled or ill-defined, some of which, as time passed, grew more pressing. Many of the new states, which became independent during the 1960s, were wary of an international legal order created before their independence which, in their view, served the interests of the industrialized nations. Pressure was increasing to develop ocean resources on a larger scale. The demand for more food coupled with new technologies resulted in an enormous increase in the harvest of fish, and the expanding search for oil at growing distances from shore created problems which demanded attention. Manganese nodules, a subject which had not been considered at the Geneva Conferences, became a matter of active speculation by the late 1960s as commercial interests began to give deep sea mining serious thought. Finally, the *Torrey Canyon* grounding made clear that modern ocean science and technology generated dangerous side-effects and that there was no adequate international machinery to deal with them.

The evolution of the law of the sea received a great stimulus in 1967 when Arvid Pardo, the Permanent Representative of Malta to the United

Nations, in a speech before the General Assembly warned against the possible appropriation of vast ocean areas by countries with the technical competence to exploit them. He proposed that the seabed and the ocean floor beyond the limits of national jurisdiction be declared the "common heritage of mankind" and that an international agency be created to assume control over this area as a trustee for all countries.

Pardo's unexpected proposal was timely in view of the increasing interest in the deep seabed and the law of the sea, and set off a chain of events leading to the third United Nations Conference on the Law of the Sea. In December of the same year, the General Assembly established an ad hoc Committee on the Peaceful Uses of the Sea-Bed and Ocean Floor Beyond the Limits of National Jurisdiction, which became known as the Sea-Bed Committee. Its main accomplishment was the draft of a Declaration of Principles Governing the Sea-Bed, which began by stating that the international area of the seabed and its resources were the "common heritage of mankind," as had been called for by Ambassador Pardo a few years earlier. The Declaration was adopted by the General Assembly in 1970 and thus became the first internationally agreed set of principles covering this vast area of ocean space.

The Sea-Bed Committee's discussions on this declaration spelled out clearly that it was impossible to consider one part of the ocean without referring to the others. Accordingly, the General Assembly broadened the Committee's mandate to cover all aspects of the law of the sea in preparation for a new conference. The enlarged committee conducted most of its work in three subcommittees: the first to prepare draft articles for the deep seabed area and its exploitation, the second to prepare a comprehensive list of issues related to the traditional law of the sea, and the third to deal with the preservation of the marine environment. Between 1970 and 1973 these subcommittees held no fewer than 469 formal meetings and produced a legacy of 161 documents. The Sea-Bed Committee completed its work in the fall of 1973, prior to the onset of the third Conference on the Law of the Sea, without producing a draft treaty, however.

7.4. THE THIRD LAW OF THE SEA CONFERENCE: 1973–1982

The third United Nations Conference on the Law of the Sea opened with a short organizational session in New York in December 1973. It began its substantive work the following year at a session in Caracas, where each of the participating states put forward its position on the whole range of law of the sea matters. In 1975, a session was held in Geneva, producing an Informal Single Negotiating Text, which, though it did not yet contain any agreed articles, indicated the direction the discussions were moving in. The next year this text was revised to reflect the progress made in some areas and became known as the Revised Single Negotiating Text. Negotiations continued on that basis until the sixth session of the conference in 1977, when an Informal Consolidated Negotiating Text was produced. It contained further changes and improvements and was reorganized to resemble more closely a draft treaty. This text underwent one more transformation to an Informal Composite Negotiating Text, which was revised again before being adopted as the new Law of the Sea Convention in 1982.

The conference was very close to reaching an agreement in 1980, but shortly after President Reagan took office in early 1981, the U.S. delegation withdrew its consent, announcing it would carefully reevaluate its position. At issue was the proposed regulation of deep sea mining and the ideology behind it, considered unacceptable by U.S. mining interests and the newly installed government. The United States' withdrawal totally disrupted the delicate consensus that had been reached, and again the split between Third World countries and industrialized nations widened. Despite some additional sessions, the adamant position of the U.S. delegation precluded a new compromise. Unwilling to wait any longer, the conference in April of 1982 adopted a formal treaty. Most Third World countries approved the convention, but the 4 negative votes and 18 abstentions included industrialized countries as well as most of the Soviet bloc, indicating that the Law of the Sea story is not quite concluded. The treaty itself, however, will enter into force one year after 60 states have ratified it. It is expected that this will occur in the mid- to late 1980s.

Back in 1973, when the Conference was just under way, it was declared, somewhat prematurely, that the opening session in New York would be followed by a working session in Caracas, after which a signing session would be held, again in New York. Few negotiators shared this optimistic belief but few could have imagined that the conference would spread over more than ten sessions, running well into the 1980s.

It should be of no surprise, however, that the third Law of the Sea Conference took much longer than originally anticipated to arrive at a generally accepted treaty. In comparison with 1958, when the Geneva Conventions were adopted, the situation had changed thoroughly: in 1958 there were 73 draft articles for consideration by 86 participating states, whereas the present conference was working its way through well over 400 draft articles—later reduced to about 300—and was attended by nearly twice as many countries. Furthermore, the General Assembly at the outset adopted an agreement that there would be no voting on any substantive matters until there was consensus on all.

The negotiations were further complicated by the variety of interest groups that emerged from the meetings of the Sea-Bed Committee. In 1958 there were essentially two such groups: nations favoring moderate coastal claims in order to preserve the freedom of the seas, and countries advocating extended coastal jurisdiction. As a result of the increase in number and complexity of ocean uses and political and economic importance, participants to the third Conference organized themselves in a much wider range of groups. There were the regional groups such as the Latin American, African, and East European groups; the issue-oriented groups such as the archipelagic states, the coastal states, and the landlocked states; and, most significantly, the north–south alliances which dominated the negotiations. With all this, the third United Nations Conference on the Law of the Sea quickly grew to become one of the largest, longest, and most complex conferences in history.

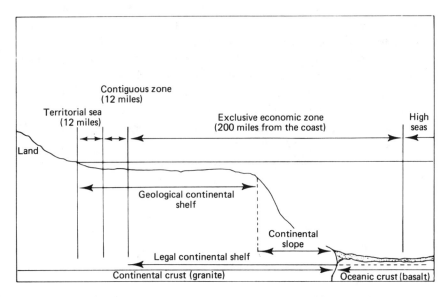

Figure 7.2. The legal division of the ocean as proposed by the third Law of the Sea Conference.

Despite the remaining differences, it is possible to discern the general outline of the emerging law of the sea clearly. Its basis is the division of the sea into new or expanded jurisdictional zones: a 12-mile territorial sea and contiguous zone, a largely expanded continental shelf, and a new 200-mile exclusive economic zone (Fig. 7.2). The major provisions of this regime are briefly discussed next.

The Territorial Sea
and the Contiguous Zone

The breadth of the territorial sea, which was one of the major unsolved questions at the first and second conferences on the law of the sea, is finally established at a maximum of 12 nautical miles, measured from the low-water line along the coast. Where the coastline is deeply indented, straight baselines can be used, as was also provided for in the 1958 Convention on the Territorial Sea and the Contiguous Zone.

The territorial sea is part of the sovereign territory of the coastal state, except that ships enjoy the right of innocent passage through it. The definition of "passage" in the new treaty is virtually the same as in the 1958 Convention, but that of "innocent passage" is much more elaborate. It still includes the basic expression that "passage is innocent as long as it is not prejudicial to the peace, good order or security of the coastal state"; in addition, however, there is a list of 12 activities specifically prohibited by passing vessels. A new article permits coastal states to prescribe traffic separation schemes in the territorial sea. In general, the provisions of the treaty represent a reasonable compromise between the interests of the coastal states and the maritime countries.

The breadth of the contiguous zone cannot extend more than 24 miles from the appropriate baseline. It remains a zone where a state may prevent and punish violations of customs, fiscal, immigration, or sanitary regulations within its territory or territorial sea.

International Straits

The adoption of a 12-mile territorial sea will greatly affect navigation in that over 100 straits would change from international waterways, where free passage is possible, to waters under coastal control, where the restraints of innocent passage apply. It should therefore be of no surprise that the major maritime states and military powers wanted to retain high seas rights—particularly free transit—in those straits that will be affected. Some coastal states, on the other hand, wanted the more restrictive territorial sea regime to apply, which was not entirely unreasonable in view of the enormous growth in world shipping and its associated dangers.

Whereas the 1958 Convention on the Territorial Sea and the Contiguous Zone contained only one sentence on the subject of straits, the new regime includes an elaborate scheme. It basically revolves around the new concept of transit passage, somewhat of a compromise between free transit and innocent passage. Transit passage is defined as "the exercise . . . of the freedom of navigation and overflight solely for the purpose of continuous and expeditious transit of the strait between one area of the high seas or an exclusive economic zone and another area of the high seas and the exclusive economic zone."

A number of articles specify the rights and duties of the strait states and the users of the straits. In accordance with the demands of the coastal states, these provisions permit strait control over fishing, pollution safety, and other potential threats. As insisted upon by the maritime nations, transit passage will not be suspended and includes submerged passage as well as the overflight of aircraft. Because of the immense importance of some of these straits, particularly the straits of Bab el Mandab, Malacca, Hormuz, and the Dardanelles, the negotiation of this new regime was an extremely sensitive issue.

Islands and Archipelagoes

The new law of the sea treaty preserves the 1958 definition of an island: "a naturally formed area of land, surrounded by water, which is above water." In 1958 that was sufficient but now there is a need to make a distinction between islands entitled to an exclusive economic zone and those that are not. Without this distinction, thousands of tiny islands such as Rockal in the Northeast Atlantic, which has a circumference of only about 100 m, could be endowed with an economic zone of 430,000 km^2. The new convention specifies therefore that rocks, which cannot sustain human habitation or economic life of their own, shall have no exclusive economic zone or continental shelf. They are entitled to territorial seas and contiguous zones, however.

The subject of archipelagoes, which was skirted in 1958, was taken up by the third conference. The treaty, in fact, provides for a regime on archipelagoes that is quite new in international law. In essence, it proposes that an archipelago, and all the water in it, forms a single entity, bound together by the sea, which has unusual economic value to its inhabitants. Baselines of archipelagoes will be drawn to connect the outermost islands rather than from the individual islands. The waters so enclosed are not internal waters, territorial waters, or high seas, but "archipelagic waters," in which ships enjoy the rights of innocent passage.

The principle proponents of this regime were, of course, the archipelagic states such as Indonesia and the Philippines. Opponents included particularly the maritime nations, for whom free passage through archipelagoes was in the same category as passage through straits. Since they had to concede on the archipelago regime, their objective was to minimize its applicability, and they were fairly successful at this. The definition of archipelagic states, for instance, specifies that they are states constituted wholly by one or more archipelagoes, excluding the coastal arachipelagoes, which are far more abundant. In addition, there are eight paragraphs detailing the drawing of the baselines; one requires a maximum ratio of water to land area of nine to one while another limits the length of baselines to 100 miles. This effectively reduces the applicability of the regime to Indonesia, the Philippines, The Fiji Islands, the Bahamas, and a few others.

The Exclusive Economic Zone

The regime of the exclusive economic zone is certainly one of the most important aspects of the third Law of the Sea Conference. While some of its elements have antecedents, the concept is new and still somewhat controversial.

Extending 200 miles from the coast, the exclusive economic zone covers about 36 percent of the total ocean surface, subtracting more than one third of what was previously considered high seas. Within this zone, the coastal state will have sweeping powers, including "sovereign rights for the purpose of exploring and exploiting, conserving and managing the natural resources, whether living or non-living, of the seabed and the subsoil and the superjacent waters, and with regard to other activities for the economic exploitation and exploration of the zone, such as the production of energy from the water, currents and winds," as well as jurisdiction with regard to the establishment and use of artificial islands, installations, and structures and marine scientific research and the protection of the marine environment.

Since the conservation of living resources in the exclusive economic zone is left solely to the coastal state, 85 to 95 percent of the world's current fish catch will fall largely under the control of coastal states. As discussed in an earlier chapter, this enclosure may not guarantee optimal utilization, but it may save certain stocks from total depletion. As a result of the demands of the distant-water fishing nations, the coastal state's control over fisheries in the economic zone is moderated somewhat by an obligation "to promote the objective of optimum utilization of the living resources," which means that coastal states should determine how much fish they can take and allow other states to harvest the surplus.

Even though the exclusive economic zone will be part of the new law of the sea when the treaty goes into effect, it remains a vague concept. In fact, more than 50 nations have already unilaterally declared economic zones, incorporating various degrees of jurisdiction, indicating that the concept will need further clarification before too long.

Finally, it is interesting to note that, though the exclusive economic zone concept was initially strongly opposed by the industrialized nations and supported by Third World countries, these positions do not necessarily reflect who will benefit. The ten countries that will gain the largest economic zones are the United States, Australia, New Zealand, Canada, the Soviet Union, Japan, Indonesia, Brazil, Mexico, and Chile. On March 10, 1983, the United States, though not a signatory of the treaty, declared an exclusive economic zone of its own by means of a presidential proclamation.

The Continental Shelf

Provisions on the continental shelf are expanded in the new treaty, reflecting the growth of its uses in the last decades and certainty of more intensive exploitation in the years ahead.

The most significant change from the 1958 Convention on the Continental Shelf is in the definition of the shelf. The 1958 definition was largely experimental and tentative, reflecting the lack of detailed knowledge of submarine geology. The new definition may be clearer, but is certainly no less confusing. It describes the shelf as the natural prolongation of a state's land territory to at least 200 miles. It may also extend beyond that point in some areas but not beyond the edge of the continental margin, which is given a technical definition in the text, intended to exclude the deep seafloor. The outer limits of the shelf may not extend beyond 350 miles from shore except in relatively shallow areas, where the line can be drawn 100 miles beyond the place where the ocean depth reaches 2500 m. Clearly, this definition leaves quite a bit of room for interpretation. In fact, during one session a paragraph was added to exclude oceanic ridges from this definition.

It can be seen that the continental shelf, as defined by the Law of the Sea negotiators, is rather different from that given by marine geologists. By including the continental slope and rise, even more is subtracted from what Ambassador Pardo once called the "common heritage of mankind." As is the case with respect to the economic zone, relatively few states will benefit from these provisions. Other states proposed that coastal states should share a part of the revenues derived from exploitation of the shelf beyond 200 miles with the international community. As a consequence, 7 percent of the revenues obtained from mining the shelf beyond this distance is supposed to be deposited in an international revenue-sharing fund.

The High Seas

That portion of the world ocean remaining after the economic zones and the archipelagic waters have been subtracted is still considered high seas. The section dealing with this issue in the new treaty is similar to the two remaining 1958 Geneva Conventions: the Convention on the High Seas and the Convention on Fishing and the Conservation of the Living Resources of the High Seas. The traditional freedoms of the high seas—navigation, overflight, fishing, and laying submarine pipelines or cables—are retained and the freedom of scientific research as well as the freedom to construct artificial islands are added. The provisions on marine transportation and on the management and conservation of fisheries are largely the same as in the 1958 Conventions. This implies that not much will be done to improve on the weaknesses of the Geneva regime, but it should be kept in mind that the high seas area has shrunk considerably from what it was in 1958.

The Deep Seabed

In the treaty the deep seabed is simply called "the Area." Officially it is defined as the "seabed and ocean floor and subsoil thereof beyond the limits of national jurisdiction," the area designated as the common heritage of mankind.

The desire to establish a regime for the development of polymetallic nodules on the deep seabed provided the impetus for the third Law of the Sea Conference, and throughout the many years of discussions it was also its most important and controversial issue.

The new treaty establishes an international agency which will manage the mineral resources of over half of the earth's surface for the benefit of all the peoples of the world. The exploitation of deep sea minerals will fall under the jurisdiction of this agency, the International Sea-Bed Authority, which will be empowered to conduct its own mining operations and to authorize private and state ventures to obtain mining rights in the Area. Under the adopted parallel system of exploitation, an area of commercially equal value will be reserved for the developing countries for every seabed area allotted to a state or private company. As indicated earlier, however, a number of industrial nations consider these provisions, particularly in what concerns the transfer of technology and the financial aspects, unacceptable.

Marine Pollution

The treaty includes a substantial section on the protection and preservation of the marine environment, unlike the 1958 Conventions, which hardly mentioned it. Whether this section will prove effective remains to be seen, however. The Conference indeed appeared more concerned with dividing up the oceans than with taking a global approach, and the pollution provisions remain very general, serving as guidelines for what states should do rather than specific criteria.

Despite these handicaps, the pollution section is quite comprehensive. Most attention is devoted to pollution caused by vessels, but there are also provisions on pollution from dumping, land-based sources, and the atmosphere. The draft text also includes clear procedures for enforcement, which will remain under the jurisdiction of the flag state on the high seas, but under certain conditions, may be assumed by the coastal state or the port state.

Marine Scientific Research

A final issue of considerable concern and disagreement at the third Law of the Sea Conference was marine scientific research. Like pollution, it was a subject that seemed of little interest to Third World countries.

They knew relatively little about it, and what they knew tended to make them suspicious. For these reasons, Third World countries successfully demanded that the coastal state must give its consent, and may impose certain conditions, before any research project may begin in its exclusive economic zone or on its continental shelf. Industrial nations, on the other hand, rightfully feared that such provisions would hamper the scientific research process, though the burden on the researcher was eased somewhat by the inclusion of the compromising doctrine of implied consent. This means that a scientist who has applied for permission to engage in a particular research project may assume that consent has been granted unless, within four months of the receipt of the application, the coastal state has specifically withheld its permission.

Chapter Eight

Conclusions

The previous chapters reviewed the extent and regulation of the most important ocean uses. In regard to fisheries, it was noted that there is a lot of food that can be obtained from the sea, though the harvest of traditional species is probably limited to a doubling, at best, of the present catch. The surplus is made up of unconventional species such as krill, lanternfish, and squid, and perhaps through the development of sea farming.

Any increases, however, will need to be accompanied by changes in the management of fisheries. This change appears to be effecting itself. Until recently, the management of fisheries was based on their designation as a common property resource, accessible to all. This was fine as long as there was a relatively small number of fishermen but led to overfishing once their number, and the catching power of their boats, increased. Management practices are moving away from that notion. Coastal states indeed are claiming jurisdiction over the fisheries up to 200 miles from their shores, thus "nationalizing" perhaps as much as 90 percent of all ocean fisheries. This move toward national control is not necessarily a guarantee of optimal use of the ocean's food resources but currently it may well be the only means for preventing further waste and misuse.

Despite the appropriation of ocean fisheries by coastal states, some type of international management regime will remain necessary since fish stocks do not discriminate between artificial boundaries and move freely from one zone to another. This is not the case with ocean minerals, which, for the most part, are static. It is this distinguishing feature that accounts for the difference in the regulation of fisheries and mineral resource exploitation. Most mineral resources are subject to national jurisdiction, whereas fisheries, until recently, were common property resources, open to all for exploitation.

The mineral resources of the continental shelf belong to the coastal state, whereas those of the deep sea belong to the world community at large. Where exactly the boundary between the continental shelf should be drawn is not yet clear since coastal states, in their efforts to obtain as much of the ocean's mineral wealth as possible, are pushing the legal limits of the shelf far beyond where geologists consider it to end.

In view of the ocean's enormous size, the amounts of marine minerals appear staggering. Many are of immense strategic and economic importance because land-based reserves are being depleted, so that marine mineral development is rapidly becoming one of the most valuable ocean activities. Even though exploitation started but a few decades ago, tremendous progress has been made in the development of recovery techniques, indicating that the exploitation of even the most remote deposits is only a matter of time and economics.

Contrary to popular opinion, waste disposal is a legitimate ocean use. Natural waters indeed have the ability for purification and there is no doubt that the oceans, as a consequence, can assimilate a share of society's wastes. Not to use this capacity would be a waste of a very valuable resource.

It appears, however, that the situation is getting out of hand. The use of the ocean as a common property resource has led not only to overfishing but in some instances to waste disposal beyond its capacity. Exactly how this is affecting the oceans is not known. In fact, our understanding of the effects of pollutants on the marine environment is still incomplete. There are instances where the effects of particular contaminants on specific species have been well documented, but these instances represent no more than a few pieces in an enormous jigsaw puzzle. Perhaps of most concern is the lack of information on the long-term effects of pollutants. It is indeed possible that small concentrations of a contaminant are not immediately noticeable but may have irreversible consequences once the concentration builds up, or accumulates, to a certain level.

Since the effects of marine pollution are international in nature, its regulation is largely a matter of international politics. In response to various incidents and problems, a substantial body of international marine pollution law has been implemented, but its record is not always impressive. Most of the treaties are not all that stringent and the enforcement procedures leave a lot to be desired. In addition, international marine pollution regulations remain to be implemented nationally, and the interpretation of these standards by different governments often vary considerably. There is no reason to believe that this will change in the near future. On the other hand, there is also no reason to believe that the oceans are "dying" as a result of pollution.

The regulation of shipping also largely takes place at the international level. Several treaties have been concluded to prevent collisions, guarantee the safety of life at sea, and establish uniformity in other practices of concern to the maritime community. As is the case with pollution prevention measures, the efficacy of this regime remains questionable at times, principally because of the lack of stringent enforcement procedures.

Fishing, navigation, mineral development, and waste disposal are by far the most important uses of the ocean, but other activities cannot be neglected. The oceans can be used as an energy source, a newly developing activity which will need to be integrated with other ocean uses. Similarly, the use of the sea for recreational and military purposes, though not discussed in this study, are becoming more important. All marine activities, traditional and new, are characterized by a growth in magnitude. As the use of the ocean intensifies, it becomes possible that the ocean is "overused" or that conflicts occur, and for these reasons there is a need for more coordination in the management of marine activities. As may have become clear from the previous chapters, ocean management is essentially vertical in nature, that is, different uses are regulated in isolation from each other. The unavoidable increase in ocean uses will require more coordination, more of a horizontal approach. The new law of the sea appears to be one step in this direction since it covers all aspects of ocean law in one treaty, though it is, of course, not much of a management tool.

Ocean management can be implemented at different levels of jurisdiction: local, national, regional, global. Since the oceans are international in nature, national measures taken by themselves would not be totally sufficient, as in most instances they can regulate only national activities. However, there is a need for additional coordination at this level. Until recently, fishing and navigation were the only ocean uses governments were concerned with, and legislation was adopted as the need for regulations or some form of control manifested itself. As more ocean uses developed, a similar piecemeal regulatory approach was taken, so that responsibilities with respect to marine activities became dispersed in numerous agencies and departments. In the United States, for instance, federal ocean programs are administered by nine departments, eight independent agencies, and some 38 agencies or subagencies. A similar fragmentary approach to ocean policy also exists in Western Europe and other developed nations. Inevitably, such fragmentation leads to overlapping responsibilities and difficulties in conflict resolution. There are too many actors and too many chains of command, and this type of regulatory structure cannot ensure intelligent ocean management.

The increase in number, magnitude, and intensity of marine activities calls for a more coordinated and streamlined approach to national ocean policies. The primary goals of such coordination should be to clearly assign responsibilities, develop an adequate conflict resolution methodology, provide for consistency as well as flexibility in ocean policies, and perhaps obtain higher level attention for ocean programs in the administrative hierarchy. This does not necessarily mean that governments should reorganize their regulatory programs, with the ocean as a central theme. In fact, in view of the many crosscutting interests in the ocean (rather than a single national interest), it is doubtful whether this would be desirable or even possible. In addition, the nature of democratic governments, which tend to generate new agencies to minister to public concerns, would certainly complicate such a task. However, a good deal of coordination can be achieved without additional institution-building. The current efforts in the United States and other developed nations toward such coordination therefore deserve total support.

There also have been calls to coordinate ocean development at the global level. While this may sound appealing, a global approach may not turn out to be very effective in view of the enormous size and diversity of the oceans. To the outsider, the ocean may seem a very homogeneous environment, but within it there is as much diversity in surroundings and life as is displayed on land. Global ocean policies, as a result, run the danger of remaining "empty shells" of sorts, much the same way a global land policy, aside from its philosophical basis, would remain relatively meaningless.

A regional approach to ocean development and regulation, in contrast, may be more effective. Already, many international agreements stress a regional course of ocean affairs, and even the new Law of the Sea Convention is replete with references to "regional and subregional organizations." A regional approach permits planning on a level that is neither too large (global) nor too confined (national or local), and therefore appears to be very useful for ocean management purposes. Of course, not all of the ocean can conveniently be divided up into regions, but some areas, many of which are intensively used, present themselves as good candidates: the North Sea, the Baltic Sea, the Mediterranean, the Caribbean, the Red Sea, or even much larger areas such as the Southwest Pacific. Regional efforts to develop these regions, already under way in several instances, are a step in the right direction.

Ocean management is one area in which the international community of nations could display some foresight. Of course, some of the political pressures may be lacking, for the ocean has not become so chaotically congested or appallingly deteriorated that the need for immediate action is evident. We should also keep in mind that nations, like individuals, are usually more interested in their own welfare than that of the world, or the sea, at large. To convince our governments that the ocean needs a more coordinated management regime will thus require continued effort from all of us.

Bibliography and Selected References

Chapter One

Broecker, W. S., *Chemical Oceanography*, Harcourt Brace Jovanovich, New York, 1974.

Drake, C. L., J. Imbrie, J. A. Knauss, and K. K. Turekian, *Oceanography*, Holt, Rinehart and Winston, New York, 1978.

Freeman, W. H., *The Ocean*, Scientific American Books, San Francisco, 1969.

Hardy, A., *The Open Sea*, Parts I and II, Collins, London, 1971.

Idyll, C. P. (ed.), *Exploring the Ocean World*, Thomas Y. Crowell, New York, 1969.

Orr, A. P. and S. M. Marshall, *The Fertile Sea*, Fishing News Books, Farnham, Surrey, 1969.

Pirie, R. G. (ed.), *Oceangraphy*, Oxford University Press, New York, 1973.

Ross, D. A., *Opportunities and Uses of the Ocean*, Springer-Verlag, New York, 1980.

Sumich, J. L., *An Introduction to the Biology of Marine Life*, Wm. C. Brown, Dubuque, Iowa, 1976.

Thurman, H. V., *Introductory Oceanography*, 3rd ed., Charles E. Merrill, Columbus, Ohio, 1981.

Van Andel, T., *Tales of an Old Ocean*, W. W. Norton, New York, 1978.

Chapter Two

Ackerfors, H., "Production of Fish and Other Animals in the Sea," *Ambio*, 1977, Vol. 6, No. 4, pp. 192-200.

Anderson, L. G., *The Economics of Fisheries Management*, Johns Hopkins University Press, Baltimore, 1977.

Bardach, J. E., *Harvest of the Sea*, Harper and Row, New York, 1968.

Bardach, J. E., J. Ryther, and W. D. McLarney, *Aquaculture*, John Wiley and Sons, New York, 1972.

Beverton, R. J. and S. J. Holt, *On the Dynamics of Exploited Fish Populations*, Her Majesty's Stationery Office, London, 1957.

Chapman, V. J., *Seaweeds and Their Uses*, Methuen, London, 1970.

Christy, F. T. and A. Scott, *The Common Wealth in Ocean Fisheries*, Johns Hopkins University Press, Baltimore, 1972.

Crutchfield, J., *The Fisheries: Problems in Resource Management*, University of Washington Press, Seattle, 1965.

Cushing, T. H., *Marine Ecology and Fisheries*, Cambridge University Press, Cambridge, 1975.

Food and Agriculture Organization (FAO), *Yearbook of Fishery Statistics*, Volume 50, Rome, 1981.

Food and Agriculture Organization, *Review of the State of World Fishery Resources*, Committee on Fisheries, Thirteenth Session, COFI/79/Inf. 4, Rome, 1979.

Food and Agriculture Organization, *Review of the State of World Fishery Resources*, Committee on Fisheries, Eleventh Session, COFI/77/Inf. 5, Rome, 1977.

Gulland, J. A., *The Management of Marine Fisheries*, Scientechnica, Bristol, 1974.

Harvey, H. W., "On the Production of Living Matter in the Sea," *Journal of the Marine Biological Association United Kingdom*, 1950, Vol. 29, pp. 97–136.

Holt, S. J., "The Food Resources of the Ocean," *Scientific American*, 1969, Volume 221, No. 3, p. 180.

Idyll, C. P., *The Sea Against Hunger*, Thomas Y. Crowell, New York, 1970.

Iversen, E. S., *Farming the Edge of the Sea*, Fishing News Books, Farnham, Surrey, 1976.

Johnston, D. M., *The International Law of Fisheries*, Yale University Press, New Haven, 1965.

Moiseev, P. A., *The Living Resources of the Ocean*, translated by the National Marine Fisheries Service, Washington, D.C., 1969.

Rothschild, B. J., *World Fisheries Policy*, University of Washington Press, Seattle, 1969.

Ryther, J. H., "Photosynthesis and Fish Production in the Sea," *Science*, 1969, Vol. 166, pp. 72–76.

Steele, J. H., *Marine Food Chains*, University of California Press, Berkeley, 1970.

Strickland, J. D. H. and T. R. Parsons, *A Practical Handbook of Sea Water Analysis*, Fisheries Research Board of Canada, 1960.

Tait, R. V. and R. S. Desanto, *Elements of Marine Ecology*, Springer-Verlag, New York, 1972.

Chapter Three

American Petroleum Institute, *Petroleum Statistics*, Washington, D.C., 1983.

Baram, M. S., D. Rice, and W. Lee, *Marine Mining of the Continental Shelf*, Ballinger, Cambridge, Massachusetts, 1978.

Cruickshank, M. J. and H. D. Hess, Marine Sand and Gravel Mining, in R. E. Osgood (ed.), *Toward a National Ocean Policy*, Johns Hopkins University Press, Baltimore, 1976.

Frank, R. A., *Deep Sea Mining and the Environment*, West Publishing Company. St. Paul, 1976.

Glasby, G. P., "Minerals from the Sea," *Endeavor*, 1979, Vol. 3, No. 2, pp. 82–85.

Horne, R. A., Recovery of Chemicals from the Sea and Desalinization, in Marine Chemistry, John Wiley and Sons, New York, 1969.

Lee, W. W. L., *Decisions in Marine Mining*, Ballinger, Cambridge, Massachusetts, 1979.

Mangone, G. J. (ed.), *The Future of Gas and Oil from the Sea*, Van Nostrand Reinhold, New York, 1983.

Mero, J. L., *The Mineral Resources of the Sea*, Elsevier, Amsterdam, 1965.

Swan, P. N., *Ocean Oil and Gas Drilling and the Law*, Oceana Publications, Dobbs Ferry, New York, 1979.

United Nations, *Manganese Nodules: Dimensions and Perspectives*, Ocean Economics and Technology Office, D. Reidel, Dordrecht, 1979.

Wenk, E., "The Physical Resources of the Ocean," *Scientific American*, 1969, Vol. 221, No. 3, 83–91.

Chapter Four

Barros, J. and D. Johnston, *The International Law of Pollution*, Free Press, New York, 1974.

Bascom, W., "The Disposal of Waste in the Ocean," *Scientific American*, 1974, Vol. 231, No. 2, pp. 17–25.

Brooks, R. R. and M. C. Rumbsky, "The Biochemistry of Trace Element Uptake by Some New Zealand Bivalves," *Limnology and Oceanography*, 1965, Vol. 10, pp. 521–527.

Cusine, D. J. and J. P. Grant, *The Impact of Marine Pollution*, Allenheld, Osmun and Co., Montclair, New Jersey, 1980.

Deese, D. A., *Nuclear Power and Radioactive Waste*, Lexington Books, Lexington, Massachusetts, 1978.

Food and Agriculture Organization, *Pollution: An International Problem for Fisheries*, World Food Problems No. 14, FAO, Rome, 1971.

Goldberg, E. D., *The Health of the Oceans*, Unesco Press, Paris, 1976.

Goldberg, E. D. (ed.), *A Guide to Marine Pollution*, Gordon and Breach, New York, 1972.

Hjalte, K., K. Lidgren, and I. Stahl, *Environmental Policy and Welfare Economics*, Cambridge University Press, Cambridge, 1977.

Hood, D. W., *Impingement of Man on the Oceans*, Wiley-Interscience, New York, 1971.

Johnston, D. M., *The Environmental Law of the Sea*, International Union for Conservation of Nature and Natural Resources, Gland, Switzerland, 1981.

Johnston, R. (ed.), *Marine Pollution*, Academic Press, New York, 1976.

Joint Group of Experts on the Scientific Aspects of Marine Pollution (GESAMP), *Pollutants in the Aquatic Environment*, FAO, Rome, 1976.

Lockwood, A. P. M., *Effects of Pollutants on Aquatic Organisms*, Cambridge University Press, Cambridge, 1975.

M'Gonigle, R. M. and M. W. Zacher, *Pollution, Politics and International Law*, University of California Press, Berkeley, 1979.

National Academy of Science, *Assessing Potential Ocean Pollutants*, Washington, D. C., 1976.

National Academy of Science, *Petroleum in the Marine Environment*, Washington, D.C., 1975.

Papadopoulou, C. and G. D. Kaias, "Trace Elements Distribution in Seven Molluscs from the Saronikos Gulf," *Acta Adriatica*, 1976, Vol. 18, pp. 367–378.

Pearson, C. S., *International Marine Environmental Policy: The Economic Dimension*, Studies in International Affairs No. 25, Johns Hopkins University Press, Baltimore, 1975.

Rice, T. R. and D. A. Wolfe, "Radioactivity—Chemical and Biological Aspects," in D. W. Hood (ed.), *Impingement of Man on the Oceans*, Wiley-Interscience, New York, 1971.

Riley, R. and R. Chester, *Introduction to Marine Chemistry*, Academic Press, New York, 1971.

United Nations Environment Programme (UNEP), *The Health of the Oceans*, UNEP Regional Seas Reports and Studies No. 16, Geneva, 1982.

Williams, J., *Introduction to Marine Pollution Control*, John Wiley and Sons, New York, 1971.

World Health Organization, *Coastal Pollution Control*, Geneva, 1976.

Chapter Five

Abrahamsson, B. J., *International Ocean Shipping: Current Concepts and Principles*, Westview Press, Boulder, Colorado, 1980.

Gold, E., *Maritime Transport*, Lexington Books, Lexington, Massachusetts, 1981.

Lloyd's Register of Shipping, *Statistical Tables 1982*, London, 1983.

Mostert, N., *Supership*, Warner Books, New York, 1975.

Ramsay, R. A., "World Trade versus the Supply of Shipping and Ships," *Marine Policy*, 1980, Vol. 4, pp. 63–66.

UNCTAD, *Review of Maritime Transport*, Geneva, 1978.

Chapter Seven

Akehurst, M., *A Modern Introduction to International Law*, George Allen and Unwin, London, 1978.

Bishop, W. W., *International Law*, Little, Brown, Boston, 1971.

Brown, E. D., *The Legal Regime of Hydrospace*, Stevens, London, 1971.

Buzan, B. G., *Seabed Politics*, Praeger, New York, 1976.

Churchill, R. et. al., *New Directions in the Law of the Sea*, Vols. 1–9, Oceana Publications, Dobbs Ferry, New York, 1973–1981.

Gamble, J. K. and G. Pontecervo, *Law of the Sea: The Emerging Regime of the Oceans*, Ballinger, Cambridge, Massachusetts, 1974.

Larson, D. D., *Major Issues of the Law of the Sea*, University of New Hampshire, Durham, 1976.

Mangone, G. J., *Law for the World Ocean*, Stevens, London, 1981.

Oda, S., *The Law of the Sea in Our Time: New Developments*, Sythoff, Leiden, 1977.

United Nations, Third Conference on the Law of the Sea, *Convention on the Law of the Sea*, 1982.

Chapter Eight

Alexander, L. *Regional Arrangements in Ocean Affairs*, Office of Naval Research, Washington, D.C., 1979.

Armstrong, J. M. and P. C. Ryner, *Ocean Management: A New Perspective*, Ann Arbor Science Publishers, Ann Arbor, Michigan, 1981.

Johnston, D. M. (ed.), *Regionalization of the Law of the Sea*, Ballinger, Cambridge, Massachusetts, 1977.

Juda, L., *Proceedings of the Symposium on Marine Regionalism*, University of Rhode Island, Kingston, 1979.

Index